The App Generation

HOWARD GARDNER
AND KATIE DAVIS

The App
Generation

HOW TODAY'S YOUTH
NAVIGATE IDENTITY, INTIMACY,
AND IMAGINATION IN
A DIGITAL WORLD

Yale UNIVERSITY PRESS
New Haven and London

Yale University Press books may be purchased in quantity for educational, business, or promotional use. For
information, please email sales.press@yale.edu (US
office) or sales@yaleup.co.uk (UK office).

Designed by Nancy Ovedovitz and set in Sabon type
by Integrated Publishing Solutions, Grand Rapids,
Michigan. Printed in the United States of America.

Library of Congress Cataloging-in-Publication Data
Gardner, Howard, 1943–
The app generation : how today's youth navigate
identity, intimacy, and imagination in a digital world /
Howard Gardner and Katie Davis.
pages   cm
Includes bibliographical references and index.
ISBN 978-0-300-19621-4 (hardback)
1. Internet and youth.   2. Youth—Social
networks.   3. Technology and youth.   4. Identity
(Psychology)   5. Creative ability in
adolescence.   6. Application software.   I. Davis,
Katie (Assistant professor).   II. Title.
HQ799.9.I58G37 2013   004.67'80835—dc23
2013017948

A catalogue record for this book is available from the
British Library. This paper meets the requirements of
ANSI/NISO Z39.48–1992 (Permanence of Paper).

10 9 8 7 6 5 4 3

For Judy Dimon
Who conceived of this project
Supported it generously
and
Always posed timely questions

Contents

viii

CONTENTS

Preface

THIS BOOK IS THE happy product of a long-standing research program, a pair of new questions, and a wonderfully fortuitous collaboration. For many years, Howard's research group at Harvard Project Zero has been studying the development in children and adolescents of cognitive mastery and an ethical orientation. Around 2006, our group began to address two new questions. Prompted by the research agenda enabled by Jonathan Fanton at the MacArthur Foundation, we began to study how the new digital media were affecting the ethical compass of young users. At roughly the same time, we began an extended, broader conversation with Judy Dimon about the ways in which young people's thought processes, personalities, imaginations, and behaviors might be affected and perhaps radically transformed by their involvement with these media.

When one launches a new project, one cannot anticipate the answers that will be forthcoming or the form that the answers will take. Katie Davis's research interests in the emerging identities of young persons stimulated her to study the blogs kept by young persons and then to undertake a dissertation on the sense of identity of young people in her home country of Bermuda. At an early point in her doctoral studies Katie joined the research group directed by Carrie James and Howard, and almost magically, a joint agenda and writing project emerged. Once we realized that today's young people could be revealingly described as the App Generation, it was clear that we needed to write this book. And given the willingness of Katie's sister Molly and of Howard's grandchild Oscar to speak with the authors, the beginning and the end of the book—covering a span of several generations—essentially wrote itself. We authors are responsible for the intervening pages.

We were fortunate to have a great deal of help with this book project. Special thanks to Carrie James, our indispensable partner in research for the past seven years. We are also grateful to our talented and dedicated research team at Harvard Project Zero, including Margaret Rundle, Celka Straughn, Margaret Weigel, and Emily Weinstein, and for more limited periods Marc Aidinoff, Zach Clark, Donna DiBartolomeo, Emma Heeschen, and Emily Kaplan. These colleagues contributed to all aspects of the research, from study design and participant recruitment to interviewing and data analysis. We have also benefited from a tremendous support staff in Howard's office: Kirsten Adam, Victoria Nichols, and Danny Mucinskas.

Katie's sister Molly was an enthusiastic and thoughtful collaborator throughout our writing, providing a valuable perspective as we reflected on the defining characteristics of the App Generation. And, while not directly involved in this project, Katie's other sister, Alaire (just one year older than Molly), was also present in our thoughts as we stitched together the three-generation narrative that runs throughout the book.

Thank you to our many interview and focus group participants, well over one hundred persons. We appreciate your willingness to spend time with us and engage our questions thoughtfully. We also wish to thank those who helped us to coordinate these interviews and focus groups, including Themis Dimon, Mary Skipper, and Shirley Veenema.

We are very grateful to Nancie Atwell, Anne Gisleson, and John and Stephanie Meyer for providing us with access to a treasure trove of art and fiction written by youth over the preceding twenty years. We were hoping to include examples of the artwork we analyzed in the pages of this book, but unfortunately we were unable to contact all of the artists to obtain their permission.

With respect to the Bermuda-based research, Katie wishes to thank the Ministry of Education and school principals for taking an interest in her research and providing access to their schools.

Others have also helped us with various aspects of the book. We thank Michael Connell, Andrew Gardner, and Justin Reich for valuable suggestions. As careful readers of the whole manuscript, we are particularly grateful to Larry Friedman, Carrie James, and Ellen Winner.

At Yale University Press, we thank particularly our editor, Eric Brandt, our manuscript editor, Laura Jones Dooley, and our publicist, Elizabeth Pelton. The whole, often challenging process of launching and steering a book project was made immeasurably easier by Hope Denekamp, Jill Kneerim, and Ike Williams of the Kneerim Williams Literary Agency.

Finally, our work wouldn't have been possible without the generous support of Judy and Jamie Dimon, as well as Jonathan Fanton, Robert Gallucci, Julie Stasch, and Connie Yowell of the MacArthur Foundation.

ONE

Introduction

A CONVERSATION

On a sunny though chilly day in March 2012, the two authors, Howard Gardner and Katie Davis, initiated a lengthy conversation with Katie's sister Molly. Ten years earlier, Katie, then in her early twenties, had begun to study with Howard, then in his late fifties. Since then they have collaborated on numerous research and writing projects, including this book. At the time of the conversation, held in Howard's office at Harvard, Molly, aged sixteen, was a junior at an independent school in New England.

Why did Howard and Katie hold and record this conversation? Since 2006, we and our fellow researchers have been examining the role technology plays in the lives of young people, often dubbed "digital natives" because they have grown up immersed in the hardware and software of the day. As researchers, we have used a variety of empirical methods to ferret out what might be the special—indeed, defining—quality of today's young people. But we came to realize that if we were to make statements, or draw conclusions, about what is *special* about digital youth today, we required key points of comparison.

Being opportunistic as well as empirical, we realized that our very own family configurations provided one comparative lens—as well as a literary device—through which to observe and chronicle the changes across the generations. Howard—on any definition of that slippery term, a "digital immigrant"— grew up in northeastern Pennsylvania in the 1950s, at a time when one could still count the number of computers in the world. Born in Canada and raised in Bermuda, Katie grew up in the late 1980s and early 1990s. During her early childhood, her Bermuda home had just one television station (CBS), which eventually expanded to three (CBS, ABC, NBC). In the mid-1990s, her parents finally installed cable at their home. Katie's access to computers was limited to once-weekly classes in the computer lab at school. In sharp contrast, Molly, who has lived in Bermuda and the United States, cannot remember a time without desktops, laptops, mobile phones, or the Internet. Wedded to her smartphone, this prototypical digital native spent her adolescence deeply immersed in Facebook, Twitter, and other social networking communities. And so our conversation across the generations—and subsequent communications among the three of us—catalyzed comparisons of three dramatically different relations to the technologies of the time.

THREE GENERATIONS, THREE TOPICS

Although our conversation ranged widely, three topics emerged as dominant and also permeate this book: our sense of personal *identity*, our *intimate* relationships to other persons, and how we exercise our creative and *imaginative* powers (hereafter, the three Is). To be sure, the nature of our species has not changed fundamentally over time. And yet we maintain that, courtesy of digital technologies, Identity, Intimacy, and Imagination have each been reconfigured significantly in recent decades. Signs of these changes can be discerned in our conversation.

As the dominant (though slightly waning) online community among both Molly's and Katie's peer groups, Facebook was a recurrent topic of discussion. Though they are Facebook friends, the sisters employ the popular social networking site in different ways. Having joined as an adult in her late twenties, Katie uses Facebook intermittently to stay connected to friends and family living across Canada, the United States, and Bermuda. For Molly, Facebook represents a far more integral part of her daily experience. Since she joined at the age of twelve, Facebook has represented a vital social context throughout her formative adolescent years.

In describing her use of and experiences on Facebook, Molly touched on a practice among some of her peers that made an immediate and striking impression on both Howard and Katie. As is the case at just about every high school, one group of students at Molly's school are considered the popular kids. The girls are attractive and the boys play varsity sports like lacrosse and soccer. Most of the varsity boys are seniors, but a few stand-out athletes are freshmen. A while back, Molly noticed that some of the senior girls who were dating senior boys started to show up on her Facebook newsfeed as being "married." Only they were married, not to their actual boyfriends, but to the freshmen boys who played on the same sports team (!).

"The popular senior girls pick out a freshman guy who is cute and popular and probably going to be really attractive when he's older. They'll kind of adopt him, and then take pictures with him, write on his wall, and flirt with him in a joking sort of way. The boys are kind of like their puppets."

Howard was surprised by this practice, noting that we typically think of girls in high school and college as being on the lookout for older men. "When I went to school, the junior and senior girls were all trying to go out with college guys."

Molly patiently explained that it's not about a real desire to date freshmen boys—after all, the girls are already dating senior boys. It's more of an initiation and reinforcement of social status. The freshmen boys are accepted into the social life of the sports team by way of the girls, who themselves use their "Facebook marriages" as further confirmation of their connection to the senior boys.

Why open with this anecdote? Because, in addition to representing an intriguing example of youth culture in a digital era, it touches on all three of the central themes in this book. With respect to personal *identity,* the Facebook marriage between freshman boy and senior girl is an act of public performance that forms part of a teen's carefully crafted online persona. Given its orientation toward an online audience, this external persona may have little connection to the teen's internal sense of self, with its associated values, beliefs, feelings, and aspirations. Yet paradoxically, if inadvertently, this electronic betrothal may contribute to an emerging sense of identity.

Issues of *intimacy* arise when we consider the new forms of social connection and interaction that have emerged with the rise of digital media. (It's hard to come up with an analog version of the Facebook marriage.) Though we identified positive aspects to these online connections in our research, the depth and authenticity of the relationships they support are sometimes questionable. Molly observed: "You never see [the senior girls and freshmen boys] hanging out as if they're good friends, like I can't see them going to each other with a problem or anything like that. But they're really good at putting on this kind of persona on Facebook of 'Everything is great and we're all friends and nothing is wrong here.'" Consider, too, that Molly, who rarely comes in contact with these teens in person, is nevertheless connected to them on Facebook.

Our final theme is *imagination,* and there's no doubt that the Facebook marriage represents an imaginative expression, if not leap. In our conversation Howard observed: "It's a bit like in mythology, the older queen picking the younger lad who has to perform for her." Of particular note is the fact that this specific act of expression is dependent on—indeed, probably inspired by—the relationship status options available on

Facebook ("married," "single," "in a relationship," "it's complicated"). In this way, the Facebook marriage illustrates how digital media give rise to new forms of imaginative expression, just as the format of this application shapes and restricts these expressions in distinct and distinctive ways.

OUR CHARACTERIZATION: THE APP GENERATION

So much for our conversation and the themes and insights contained therein. We believe that one can find similar trends and manifestations across other areas that we might have surveyed—say, how one thinks about education or childrearing, religion or politics, work or play, personal morality or ethics at the workplace. (We consider some of these spheres at the end of the book.) The digital media leave few areas untouched—and their influences going forward promise to be equally dramatic and equally difficult to anticipate.

Yet we've become convinced that a single characterization best captures what is special about the changes digital media have wrought to this point. We capture this insight with the epithet the "App Generation." An "app" or "application" is a software program, often designed to run on a mobile device, that allows the user to carry out one or more operations. As captured in the photograph here, apps can be narrow or broad, simple or grand, and in either case are tightly controlled by the individual or organization that designed the app. Apps can access tunes or the *New York Times,* enable games or prayers,

answer questions or raise new ones. Crucially, they are fast, on demand, just in time. You might think of them as shortcuts: they take you straight to what you're looking for, no need to perform a web search or, if determinedly old-fashioned, a search through your own memory.

It's our argument that young people growing up in our time are not only immersed in apps: they've come to think of the world as an ensemble of apps, to see their lives as a string of ordered apps, or perhaps, in many cases, a single, extended, cradle-to-grave app. (We've labeled this overarching app a "super-app.") Whatever human beings might want should be provided by apps; if the desired app doesn't yet exist, it should be devised right away by someone (perhaps the seeker); and if no app can be imagined or devised, then the desire (or fear or conundrum) simply does not (or at least should not) matter.

Let's consider a familiar task of life and how an ensemble of apps has increasingly taken over how we accomplish it: finding your way from point A to point B. A century ago, if one wished to make one's way from Harvard Square in Cambridge, Massachusetts, to Boston's North End, one had a few options. One could ask a friend or passerby for written or oral directions, rely on one's memory of a previous journey, or look at a map of the greater Boston area and plan one's path by foot or some other mode of transportation. At the extremes, there were other choices: one could embark on a random walk (risking the possibility of never getting there). Or, like the proverbial Charlie from the Kingston Trio's famous song who got stuck on the Boston subway system and "never returned," one could take the MBTA. Or one could ask an organization (in later years, the American Automobile Association) for a TripTik—a fool-proof, block-by-block itinerary.

While some readers will remember these times, to contemporary consciousness they seem hopelessly old-fashioned. In recent years, many of us have in our hands or in our vehicles a device that informs us of our precise position in space, directs us from that position to our desired location, and, if for any reason we deviate from the preferred route, adjusts its directions accordingly. For all practical purposes, such GPS systems are apps that remove uncertainty from our journey. Indeed, you might decide to use your smartphone as a navigation system by calling up Google Maps. Such apps not only give us extremely detailed maps of locations; drawing on our known and inferred preferences and the reviews of other users, they inform us about options every step of the way, such as nearby restaurants, cafés, or points of interest. We can say that these apps allow error-free navigation even as they seek to satisfy all of our possible needs and desires en route.

With respect to a life with foolproof navigational aids, after Howard delivered a talk on education to a college audience, a bright and somewhat aggressive student brandishing his smartphone approached Howard. Flashing a grin, he said, "In the future, why will we need school? After all, the answers to all questions are—or soon will be—contained in this smartphone." Howard reflected for a moment and then responded, "Yes, the answers to all questions . . . except the important ones." A world permeated by apps can in many ways be a wondrous one; and yet, we must ask whether all of life is—or should be—simply a collection of apps or one great, overarching super-app.

Apps are great if they take care of ordinary stuff and thereby free us to explore new paths, form deeper relationships, ponder the biggest mysteries of life, forge a unique and meaningful identity. But if apps merely turn us into more skilled couch potatoes who do not think for ourselves, or pose new questions, or develop significant relationships, or fashion an appropriate, rounded, and continually evolving sense of self, then the apps simply line the road to serfdom, psychologically speaking. One can get from Harvard Square to the North End with one's eyes wide open or one's eyes shut tight. In what follows, we attempt to capture this contrast neologistically: apps that allow or encourage us to pursue new possibilities are *app-enabling*. In contrast, when we allow apps to restrict or determine our procedures, choices, and goals, we become *app-dependent*.

In informal terms, we've introduced the problematic of this book and hinted at the answers we detail in the pages that follow. But we are hardly the first to have attempted a description of the current generation of young persons, nor are we alone in seeking to link the profile of today's youth to the influence of digital media. Indeed, hardly a day goes by without some pundit singing the praises or lamenting the costs of a life dominated by digital devices. And hardly a fortnight goes by without a major essay or book on the topic. Before plunging into the details of our study, we owe the reader an explanation of what is special about our endeavor and the book it has spawned.

Although some of the current thinking and writing about digital youth is notable, the ratio of claims made to data gathered and analyzed systematically is embarrassingly, indeed unacceptably, high. We have attempted to redress this imbalance. Over the past five years, our research team at Harvard has carried out a number of convergent studies on the nature of today's youth. Using a variety of methods, we have sought to understand to what extent, and in which ways, the youth of today may differ from their predecessors.

To begin with, we've observed young people, talked with them, eavesdropped (with permission!) on conversations dedicated to bland topics like "today's youth" or stimulated by more provocative conversation-openers like "What do we owe to our parents and for what should we blame them?" Some of these conversations have been recorded, others reconstructed based on notes we've taken.

In formal work, guided by protocols, we've conducted systematic interviews with approximately 150 young people living in the New England area and a smaller sample in Bermuda. The New England interviews were conducted between 2008 and 2010 as part of a project examining the ethical dimensions of young people's digital media activities. For this project, we spoke with youth spanning middle schoolers up to recent college graduates about their experiences with digital media, including any "thorny" situations they'd encountered online. We also interviewed twenty girls who had been blogging throughout their middle and high school years in an online journaling community called LiveJournal. The remaining youth interviews were conducted in Bermuda with students ranging from the eighth through twelfth grades. In our interviews, we have secured much information about how young people think of digital media, how they make use of them, and what they see as the advantages and the limitations of the panoply of devices at their fingertips.

To supplement our studies with young people, we have carried out an ambitious, complementary program of research with knowledgeable adults. We constituted seven focus groups, each composed of six to ten adults who had worked with young people over at least a twenty-year period—spanning the predigital to the hyperdigital era. Each focus group assembled adults who had a particular form of contact with young people. Specifically, there were focus groups composed of psychoanalysts; psychologists and other mental health workers; camp directors and longtime counselors; religious leaders; arts educators; and classroom teachers and after-school educators who worked primarily with youth living in low-income neighborhoods. In addition, we carried out forty interviews, many extending over two hours, with high school teachers who had worked with young people over at least two decades. Each focus group and interview was recorded, documented, and analyzed.

Last, in what we believe is a unique line of research, we have compared the artistic productions of young people gathered over a twenty-year period using depositories of student work that had been accumulated continuously throughout the two decades. We chose to look at two bodies of work— student writing and student graphic productions—and discern how the work changed over this period. Our findings, detailed in our discussion of the imaginative powers of young people, revealed the importance of the particular medium of expression chosen by the young artists.

So much for our methods: technical details are provided in

the appendix. Another downside of most current discussions of young people is that they are lamentably anachronistic—that is, they lack careful attention to the various contexts within which a discussion of today's young people needs to be considered.

Accordingly, in the next chapters, we provide two disciplinary contexts within which to locate the young people of today. The first context is *technological*. If we are to claim that the youth of today are defined by the technologies they favor, we need to consider how, in earlier times, technologies—ranging from hand tools to telephones—may have affected or even defined human beings, human nature, and human consciousness. This discussion invites us to distinguish among tools, machines, and the information-rich media of the past century and to consider how digital media may represent a quantum leap in power and influence.

The second context is deliberately interdisciplinary. We ask, "What do we mean when we speak of a generation?" For most of human history, generations have been defined biologically—the time from an individual's birth to when that individual becomes (or could become) a parent. In recent centuries, generations have been increasingly defined by sociological considerations. The defining characteristics of a generation echo the dominant events of the time, be they military (the Great War), political (the assassination of a leader), economic (the Great Depression), or cultural (the Lost Generation of the 1920s, the Beat Generation of the 1950s). We propose that, going forward, generations may be defined by their dominant technologies, with the length of the generation dependent on the longevity of a particular technological innovation.

Throughout our discussion, we keep our eyes on how young people have acted—as well as how they have been characterized and defined by their elders. At the same time, we maintain a sharp focus on the events of the past half century—specifically, the events that defined the spaces in which Howard, Katie, and Molly have each grown up and have helped to fashion the identity, intimacy, and imagination of the three of us, and of our peer groups. As it happens, two books published in 1950—*The Lonely Crowd,* by the sociologist David Riesman and his colleagues, and *Childhood and Society,* by the psychoanalyst Erik Erikson—provide apt contexts for this transgenerational comparison.

In such a wide-ranging undertaking, with both empirical variety and disciplinary reach, we (as well as our readers) welcome a viable and dependable throughline. This throughline is provided by our characterization of today's young people as the App Generation. Whether we are unpacking the technological or generational contexts, or reviewing our various empirical studies, we focus on how the availability, proliferation, and power of apps mark the young persons of our time as different and special—indeed, how their consciousness is formed by immersion in a sea of apps. Fittingly, in the concluding chapter, we consider the effect of an "app milieu" on a range of human activities and aspirations. More grandly, we ponder the questions, "What might life in an 'app world' signal for the future of the species and the planet?"

TWO

Talk about Technology

THE FIRST TECHNOLOGIES ARE built into our species' hard-ware and software. Stroke the side of a newborn's foot and the toes will spread; make a sudden loud sound and the infant will startle; smile at a three-month-old and the baby will smile back. No instruction is necessary.

Externally invented technologies have been with us for many thousands of years, and they are equally a part of human development. One can tickle with a brush as well as with the hand; the loud sound can come from a percussion instrument or a foghorn; and the infant can smile at a doll or a mobile. Nor need the young child be a passive reactor. Within the first year or so of life, the child can shake a rattle, search for a hidden phone, even drag a computer mouse and behold an object skipping across a screen . . . or, in the manner of the only slightly fanciful cartoon reproduced here, transfer funds from one account to another.

Whether part of each of our bodies, or devised by human

"It's very important that you try very, very hard to remember where you electronically transferred Mommy and Daddy's assets."

Michael Maslin / The New Yorker Collection.

hands over the years, technologies provide a principal means by which we carry out actions from the time of birth to the time of death—or at least until senescence appears. Many of our greatest human achievements are due to technologies devised by humans—think of clocks, the spinning wheel, the steam engine, rocket ships. Many of our most frightening achievements are also due to technologies devised by human beings—think of bows and arrows, rifles, nuclear weapons, rocket ships (again), or, most recently, the drones with which battles in remote sites are increasingly being waged.

FOUR SPHERES TO KEEP IN MIND

In our focus on apps, we are examining a preeminent technology of our time. But in discussing apps and "The App Generation," we will inevitably touch on four different perspectives or spheres, each with its own terminology and vocabulary. These perspectives are often confused or conflated in writing—and, indeed, in *thinking* about the components and forces that characterize our fast-changing era. As much as possible, with the aim of avoiding both preciosity and pedantry, we will try to make clear on which perspective we are focusing.

- Tools and machines: *technology* in the traditional sense (ax, steam engine), typically built out of wood, metal, plastic, or other available materials;
- *Information* that can be transmitted via our own bodies or by manmade technologies of various sorts (news, entertainment, maps, encyclopedia entries);
- Information transmitted by a particular machine or tool (the television set that conveys local or international news constitutes a medium of communication; so, too, the geographical information presented on a Google or Yahoo! map)—in referring to these instances, we will use the terms *medium* and its plural, *media;* and
- *Human psychology* (sensing, attending, categorizing, deciding, acting, other processes of the mind).

So, to be concrete, suppose we are dealing with options that allow a user to find out about different restaurants in a neighborhood, such as the North End of Boston.

- *Technology* is the particular smartphone or hardware that is accessed by the user, in this case, a teenager who wants to meet a group of friends for a meal;
- *Information* is the particular set of categories of food and location that can be captured in many ways;
- *Medium* is how this information can be presented in a particular app; at the time that this book went to press, Yelp and Google Maps would be popular choices, but essentially the same information could also be written down, presented in a map, or be part of another app, say one devoted to Healthy Foods; and
- *Human psychology* entails the use of hands, eyes, ears; the attention span needed to assimilate and process the information; the decision made about where to go, with whom, and for what purposes; and reflections on "how it went."

It's not uncommon to speak of technologies as changing human nature—or at least human thought and action (what we've just labeled "human psychology")—in fundamental ways. Books have been written about the changes wrought by clocks, steam engines, nuclear weapons—indeed, famously, by "guns, germs, and steel." For the American cultural critic Lewis Mumford, the technologies of the twentieth century have increasingly come to control the options available to us, making us more and more like cogs that allow our machinery

Charlie Chaplin, *Modern Times* (1936). Film still © Roy Export S.A.S.
Scan Courtesy Cineteca di Bologna.

to operate as it has been designed (initially, by human beings)
to operate.[1] We create factory machines to automate work,
and they end up converting us into automata—reminiscent
of the hurried and harried assembly-line worker in Charlie
Chaplin's *Modern Times*.

Jacques Ellul, a French contemporary of Mumford's, puts
forth a far more chilling portrait.[2] He recognizes the importance throughout history of tools—usually handheld creations that allow individual farmers or craftsmen to accomplish
daily tasks with greater efficiency. He distinguishes such tools
from machines—more elaborate devices that operate primar-

ily on their own (beyond hand-holding) and make possible mass production by assembly-line workers. But in Ellul's view, it is naive to think of machines and tools as merely coming to dominate our lives. As he sees it, such technological artifacts usher in a fundamental change in human psychology: a way of thinking in which every aspect of our lives has to be rationalized as much as possible, measured to the nth degree, rank ordered in terms of ever greater efficiency (or some other readily quantified dimension like speed or number of "hits"). Whatever contributes to these trends must be pursued; anything that gets in the way will—indeed, *has to be*—scuttled. We end up with a species that is well embarked on a single, unidirectional, unwavering march toward a totally technological milieu.

While Mumford might see apps as sapping individual agency, Ellul would see them as symptoms of an all-encompassing weltanschauung, or worldview. Human beings only too willingly accept the premises of technology—that efficiency, automaticity, impersonality can and should trump individual goals, will, faith. Put succinctly, technology re-creates human psychology.

Our interest here is centered on specific technologies (mechanical devices) that enable communication of information (hence, in our term, on particular media). Few doubt that the invention of writing, in the millennia before the birth of Christ, brought about fundamental change in human thought and expression. Socrates thought that writing would vitiate human memory, but in fact it enabled philosophical and scientific thought. There is similar consensus that the invention of the printing press 650 years ago was epochal. Gutenberg's machine undermined religious authoritarianism even as it laid the groundwork for mass education.

In the last century, in developed or developing countries, the technologies of the body, the tool chest, the factory, the weapons arsenal have been rapidly expanded and often supplanted by powerful *media of communication*. First the telegraph, then the telephone, then radio and television are objects to be touched and manipulated, entities from which to receive messages and, in the case of the telegraph (at least for those fluent in Morse or some other code) and the telephone (for anyone willing to speak up), to transmit messages as well. The specific technologies/machines are important, to be sure, but they often become inaudible and invisible, part of the background scenery—like the television sets hoisted above nearly every restaurant bar.

While some of us are inclined to think of these communication media as "mere tools," they can have a transformative effect. Replacing sea or land transportation that takes days or even weeks, telegraphs allowed the transmission of important news in minutes. Telephones permit us to communicate almost instantly with people—known to us or not—close by or at great distances. Radio and television allow us *direct access* to what is going on all over the world—news, finance, sports—as it unfolds, and provide an endless diet of entertainment, ranging from slapstick comedy to soap operas to serious drama. In December 1936, one could actually listen to King Edward VIII abdicate his throne; two years later, one could hear the cheers throughout Yankee Stadium as the black American boxer Joe Louis knocked out his German heavyweight opponent, Max Schmeling, in one round. Movies create stars and stories and scandals that are recognized around the globe.

While Lewis Mumford and Jacques Ellul reflected critically on the full range of tools and machines, Canadian scholar Marshall McLuhan focused sharply on the mass media of communication that dominated the twentieth century.[3] He compared the world of radio and television with the earlier "Gutenberg galaxy," the world of books and print, which literate people absorbed in linear order, at their own rate, with their often idiosyncratic system of markings of content. As McLuhan saw it, each medium—which he viewed as an extension of human sensory organs—alters the relation of the individual to the surrounding world. Absorbed by the eye, one saccade at a time, print pushed toward individuality, self-direction; in contrast, the electronic media of the twentieth century catalyzed a shared, ambient tribal consciousness. Media differed from one another in the extent to which they invited, or even permitted, active participation on the part of a member of the audience: "cool" invited or at least enabled participation, "hot" catalyzed passivity and dependence. In effect anticipating the Internet and the World Wide Web, McLuhan wrote about the emergence of a global village, in which humans around the planet increasingly partook, often simultaneously, of a single, generalized consciousness. It has been said that in 1997, within two days of its occurrence, 98 percent of the world (except young children) knew about the death, in a car accident, of Britain's charismatic Princess Diana.

Despite his prescience, McLuhan essentially lived and wrote in the middle of the twentieth century—an age of mass electronic media (the world of Howard's youth), rather than one of digital hegemony. Only in the succeeding decades (the era of Molly's youth) has our world come to be dominated by computers within the grasp of human beings almost everywhere. Desktops, laptops, smartphones, tablets, and other digital technologies do more than allow us to contact any and all individuals around the globe. In sharp distinction to the mass media of the last century, they are intensively personal and invite activity on the part of the user: *personal* in the sense that the individual user is (in contrast to radio and television) increasingly in control of what is received and when it is received; inviting *activity* in the sense that (again, in contrast to radio and television) it is easy and straightforward to transmit content as well as to receive it and (in contrast to the telephone or the radio) in that digital devices can readily and actively involve the visual and tactile senses, as well as the auditory. No longer do we simply receive messages from designated spots (and producers) around the world; we are now in a unique position to transmit our own messages in a variety of formats to anyone with access to digital devices.

This transition is captured vividly by the appearance of the first personal computers in the late 1970s and early 1980s (Apple 1 appeared in 1976, the Apple Macintosh [soon abbreviated to "Mac"] appeared in 1984; as if prophetically, Marshall McLuhan died in 1980). For the first time in human history it became possible for ordinary persons, not just scientists or military personnel, to have at their fingertips (indeed, at the touch of a mouse) technology that connected them instantly with the rest of the world. Anyone with a personal computer could contact other persons, create literary or graphic material or musical materials, and receive similar kinds of materials from anyone else (individual, group, corporation) that had access to comparable software and hardware. And all this communication occurred courtesy of a single elegantly designed, seductively responsive machine. While the technologies and media have changed enormously in the succeeding years, thanks in large measure to Steve Jobs and Apple Inc., we may never again experience the transcendent experience of that moment. We are reminded of Wordsworth's poetic line: "Bliss was it in that dawn to be alive, but to be young was very heaven."[4]

APPS AND HABITS

Enter apps. Only a small (albeit growing) minority at any age can write code and thereby create our own programs and procedures. Most of what we accomplish online is a result of procedures that have been created by others, with their options delimited in various ways for various purposes. And so we encounter the paradox of *action* and *restriction*. The feeling of instituting and implementing an app is active; and yet the moves enabled by each are restricted to a greater or lesser extent (for paid apps, even access is restricted). It has been said that, in this respect, an app resembles "a gated community."[5] Restrictions can either be constricting (in our terms, dictating an app-dependent frame of mind), limiting the options available, or they can constitute a challenge—asking us what we can accomplish, despite these restrictions. They can also stimulate us to create a new application or even a new *kind* of application, thereby altering our environment so that it becomes app-enabling. (Of course, even if we do create a new app, Apple may not accept it into its App Store!) In Mumford's terms, the issue is whether we will control the technologies or whether the technologies will control us. In Ellul's terms, will applications reinforce the move toward the all-encompassing technological worldview, or will they launch new forms of expression and understanding? In McLuhan's terms, are the apps simply the newest medium, with its characteristic sensory ratio? Or do they constitute an ingenious blend of the range of electronic and digital media and open up a new chapter of human psychological possibilities?

CONTRASTING PSYCHOLOGIES

When we think of a child or an adult employing an app, we shift our perspective from technology to psychology—from the machine or the medium to the human users. In the beginning, infants are characterized by an ensemble of reflexes—

sucking, looking, grasping, startling. But these reflexes are soon supplemented and eventually supplanted by a wide range of actions that reflect a congeries of factors: the maturation of the nervous system; the specific contours of the physical environment and the culture in which the child is growing; and the pattern of intrinsic and extrinsic rewards that attend these actions. We are the species par excellence of new experiences, new actions, and new reactions. And yet we could hardly advance beyond the reflex stage unless we were gifted at creating and, whenever possible, relying on new actions that evolve into long-term habits.

As is often the case in the discipline that he helped to found, psychologist William James memorably captured this phenomenon. In his phrase, habits are "the enormous flywheel of civilization." Less poetically, they make possible the rhythm of daily life as well as the potential for human progress or human regression. Indeed, the range of habits is as broad as the array of human actions and technologies. We can acquire the habit of sucking a thumb, reciting a prayer, or solving differential equations. While we are young, habits are readily acquired and rather readily altered. As James quipped, "Could the young but realize how soon they will become mere walking bundles of habits, they would give more heed to their conduct while in the plastic state. We are spinning our own fates, good or evil, and never to be undone."[6] Indeed, the world over, child rearing is an effort to instill habits that are productive—cleaning up one's mess, practicing an instrument—while attempting to extinguish those that are unproductive, harmful to self, harmful to others. We do not want our children to daydream during lessons, cross the street without looking both ways, lash out at someone when they become frustrated.

Let's remain in the world of psychology, a world in which Katie and Howard spend many working hours. We begin with a study that, we believe, deserves to be as well known as the famous "marshmallow experiment"—the one that documents the extent to which future SAT scores can be predicted from a toddler's capacity to withhold gratification when in the presence of an inviting sweet.[7] Psychologist Elizabeth Bonawitz and colleagues exposed toddlers to a toy. In one condition, which we'll call the "teaching condition," a knowledgeable adult demonstrated how to use the toy. Specifically, she showed that when one yanked a yellow tube, a squeaky sound resulted. In a second condition, which we'll call the "exploring condition," an apparently naive adult introduced the toy and, apparently by accident, executed an action that yielded the squeaky sound. Thereafter, toddlers were given the chance to play with the toy as they liked. In the teaching condition, the toddlers essentially repeated the use modeled by the adult, and that was that. But in the exploring condition, toddlers spent far more time with the toy and tried to ferret out various possible uses, extending well beyond those accidentally displayed by the naive adult (the same results were obtained with other nonteaching "control" conditions).[8]

With a perhaps permissible degree of hyperbole, we suggest that, on the basis of this one experimental result, one can build entire psychologies and complete educational philosophies. The teaching condition epitomized the psychological approach called "behaviorism." In this brand of psychology, made most famous by B. F. Skinner with his Ping-Pong–playing pigeons and infants raised in Skinner boxes, human psychology consists simply of the organism's reactions to stimuli presented by others.[9] If a behavior is rewarded, it is repeated; if it is not rewarded, it is sooner or later extinguished. In the less happy instance, humans learn by random exploration, until they happen to find a rewarding condition, in which case they persist in that situation. In the happier instance, desired behaviors are modeled and imitated.

The rival brand of psychology, which came into prominence during Howard's own professional lifetime, is called cognitivism or constructivism.[10] On this view, skills and knowledge are constructed on the basis of the individual's own active explorations of the environment. Rewards supplied by others are fine, but the most important activities are ones that are intrinsically rewarding—based on one's own discovered pleasures as one explores the world. Imitations and modeling are possible and may be helpful; but unless one makes knowledge on one's own, it remains both tenuous and tentative.

You can easily see the integral link between these psychological theories and their associated educational regimens. Behaviorists favor the most tightly structured learning environments—generously termed "well-structured curricula and tests," less kindly termed "drill and kill." In sharp contrast, constructivists call for rich and inviting problems and puzzles, which will engage curiosity and catalyze extensive exploration—with, at most, the "guide on the side," rather than the "sage on the stage." On the constructivist view, the best way to educate is to provide inviting materials and get out of the way.

Both behaviorists and constructivists recognize the importance of habits. For behaviorists they are simply the way that we all lead our lives—as Skinner pugnaciously and unsentimentally put it, lives "beyond freedom and dignity."[11] For constructivists, habits are a mixed blessing—needed to move on, yet possible barriers to continuing growth. To borrow another oft-quoted psychological phrase, habits can make it more difficult for us to proceed "beyond the information given." In our own terms, we may think of habits as potentially making us dependent on certain conditions or as enabling us, freeing us to do new and potentially important things.

The advent of the digital world introduces a bevy of potential new habits. These start with the simple inclination to use—or to spurn—a particular technology. In the time of Howard's childhood, one could either gab endlessly on the telephone or, as his parents urged, "take it off the hook." In our time, one can either keep one's smartphone by one's side night as well as day, put it aside during periods of relaxation or study, or take the unusual step of "burying it for the summer," as is now mandated in some summer camps.

(Of course, "mandated" does not mean "guaranteed" or "enforced." At one summer camp about which we learned in our study, campers engaged in an elaborate ritual in which each smartphone was placed into a receptacle, to be returned at the end of the camp session. Yet, unknown to the staff, some of the parents had hidden a second smartphone inside the campers' belongings, so that child and parent could remain in touch at will. Habits can die hard not only for digital natives but also for digital immigrants—the parents.)

The decision to use—or not use—one's devices is just the beginning. One's digital habits can range from mindless repetition of a few regular "moves" to a flexible orchestration and deployment of several disparate activities. As documented by ethnographer Mimi Ito and her colleagues, most young people in America use their devices simply to "hang out"; that is, they check in regularly with their friends to see what is going on, exchange brief greetings, plan future encounters ("Hey, what's up?" optional addressee, "Dude!").[12] This use is habitual in the least imaginative sense. A minority of young people "mess around"; that is, they seek more actively to explore a particular activity, perhaps learning some steps in Photoshop or transmitting amusing video clips to a group of friends and soliciting their reactions. In this case, the "messers" are enjoying and seeking a modest expansion of their knowledge or skills, either by themselves or in exchange with others. And perhaps 10 percent of youth actively "geek out"; they spend a significant amount of time, daily or even on the hour, developing a work or play or art skill to a high degree, seeking ever greater mastery, frequently in the company of others who share their passion. Of course, each of these groups makes use of existing apps, but only in the latter case is there an active attempt to stretch the app to its limit or, in the extreme, to create and disseminate new apps or to venture where no app has yet traveled. In the psychological terms just introduced, we can see apps either as the latest shaping technology in the repertoire of the behavioral psychologist or educator, or as a technological lever for inducing the kind of exploration endorsed by the constructivist psychologist or educator.

We can get a sense of these contrasting stances by considering two widely used apps.

With respect to Wikipedia (which is available as an app on smartphones and tablets), the minimalist approach is simply to copy or paraphrase an entry as part of a homework assignment. In contrast, should one use the Wikipedia entry as a point of departure for further research, or even edit an earlier entry in light of the dividends of such research, one enters the cohort of the geeks. Taking an example from the graphic realm, one can use one's phone's video capabilities to create the one millionth video of a cute cat, or, geek style, one could sketch out and then produce an original video about an issue on which one has strong feelings and circulate it to as wide an audience as possible.

As we step back from this foray into technologies and psychologies, let's frame the options. From the point of view of technologies themselves, we can distinguish two categories: those apps that, like Bonawitz's teaching condition, seem to dictate one's course of action, hence inculcating dependence; and those apps that, like the exploring condition, appear to open up one's possible courses of action, thereby enabling the user. From the point of view of human psychology, we can again distinguish two categories: those individuals (and their elders) who are willing or even eager to become dependent; and those individuals (and their elders) who spurn the habitual and search for conditions that are enabling. Of course, many apps will straddle these categories; and many human beings oscillate, comfortably or uncomfortably, between dependence and independence. But at least at the extremes, the contrasts are stark and important.

POSSIBILITIES AND PROBABILITIES

In light of the alternative scenarios, let's return to the three topics we're investigating here.

With respect to *identity formation:* Apps can short-circuit identity formation, pushing you into being someone else's avatar (that of your parents, your friends, or one formulated by some app producer)—or, by foregrounding various options, they can allow you to approach identity formation more deliberately, holistically, thoughtfully. You may end up with a stronger and more powerful identity, or you may succumb to a prepackaged identity or to endless role diffusion.

With respect to *intimacy:* Apps can facilitate superficial ties, discourage face-to-face confrontations and interactions, suggest that all human relations can be classified if not predetermined in advance—or they can expose you to a much wider world, provide novel ways of relating to people, while not preventing you from shutting off the devices as warranted—and that puts YOU in charge of the APPS rather than vice versa. You may end up with deeper and longer-lasting relations to others, or with a superficial stance better described as cool, isolated, or transactional.

With respect to *imagination:* Apps can make you lazy, discourage the development of new skills, limit you to mimicry or tiny trivial tweaks or tweets—or they can open up whole new worlds for imagining, creating, producing, remixing, even forging new identities and enabling rich forms of intimacy.

The Flywheel can liberate you or keep you going around in circles.

As for the probability of these various alternatives, heated debate already exists in the writings of the digerati. On the one side we find unabashed enthusiasts of the digital world. In the view of experts like danah boyd, Cathy Davidson, Henry Jenkins, Clay Shirky, and David Weinberger, the digital media hold the promise of ushering in an age of unparalleled democratic participation, mastery of diverse skills and areas of knowledge, and creative expression in various media, singularly or orchestrally.[13] As they see it, for perhaps the first time in human history, it is possible for each of us to have access to the full range of information and opinions, to inform ourselves, to make judicious decisions about or our own lives, to form links with others who want to achieve similar goals—be they political, economic, or cultural—and to benefit from the enhanced intelligence and wisdom enabled by a vast multinetworked system. On this perspective, a world replete with apps is a world in which endless options arise, with at least the majority tilted in positive, world-building, personally fulfilling directions. It's a constructivist's dream.

Others are less sanguine. Nicholas Carr claims that, with their speed and brevity, the digital media encourage superficial thinking, thereby thwarting the sustained reading and reflection enabled broadly by the Gutenberg era.[14] Raising the stakes, Mark Bauerlein invokes the inflammatory epithet "the dumbest generation."[15] Cass Sunstein fears that the digital media encourage us to consort with like-minded persons; far from exposing us to a range of opinions and broadening our horizons, the media enable—or, more perniciously, dictate—the creation of intellectual and artistic silos or echo chambers.[16] Sherry Turkle worries about an increasing sense of isolation and the demise of open, exploratory conversations, while Jaron Lanier laments threats to our poetic, musical, and artistic souls.[17] On this perspective, an app-filled world brings about dependence on the particulars of each currently popular app, and a general expectation that one's future—indeed, the future itself—will be dictated by the technological options of the time. It's a constructivist's nightmare.

Drawn from diverse sources, our data speak to these debates. As we argue in what follows, the emergence of an "app" culture allows individuals readily to enact superficial aspects of identity, intimacy, and imagination. Whether we can go on to fulfill our full potential in these spheres, to take advantage of apps ("enabling") without being programmed by them ("dependent"), remains a formidable challenge.

THREE

Unpacking the Generations: From Biology to Culture to Technology

EVER SINCE HUMANS BECAME aware that organisms are re-produced, it has been possible to think of life in terms of generations. Literally, any person, nonhuman animal, or plant is the product of the preceding (parental) generation and in turn has the potential to spawn the succeeding (or offspring) generation. (For present purposes, we'll ignore the hapless mule.) Those of us raised in the Judeo-Christian traditions probably first encountered the formal idea of generations in the Bible—through the endless list of "begats." And of course, any young person who strays beyond the nuclear family encounters individuals of older generations—aunts, uncles, grandparents, the odd great-grandparent of a far-removed generation, as well as members of one's own generation—cousins of various stripes and degrees of separation. Given the traditional generational spans, Katie could easily be Howard's daughter, Molly his granddaughter.

Bearing in mind considerations of conception, calendars,

and consciousness, what is a generation and how long does it last? In the classical era and in biblical times, the definition of a generation seems to have been straightforward: a generation spanned the period from one's birth to the time that one had offspring, at which point the offspring's own generational clock began to tick until it had children (more technically, the time from the birth of a woman to the birth of her first child). We need to bear in mind that life spans in those days were much shorter—if we exclude the biblical patriarch Methuselah, who purportedly reached the mind-boggling age of 969 years—and that adulthood began in effect at or shortly after the onset of puberty. (Some authorities suggest that, like the average life span, the length of a familial generation has virtually doubled in the course of recorded history—from fourteen to fifteen years to twenty-eight to thirty years.) We should also note that if there were informal characterizations of generations—say, the generation that went on the First Crusade to the Holy Land or the generation that lived when the Americas were discovered—these descriptors would have been less likely to be widely known and certainly less bandied about in a pre–mass media age.

Related to the definition of a generation is the way in which periods of life have been described. As far back as Homeric Greece, when the riddle of the Sphinx was initially posed, there has certainly been a recognition of the difference between the young child (four legs), the mature adult (two legs), and the old person (three legs)—though that older person might well be forty or fifty, rather than the biblical "three score and ten." Shakespeare memorably delineated the seven stages of man. We cannot know precisely which chronological ages the Bard had in mind nor the duration of each stage. But we can assume that the period of *youth* came after that of the *infant* and the *schoolboy*, and before the latter three ages of *justice, early old age,* and *second childhood.* More concretely, using the descriptors favored by Shakespeare, we can point to *lovers*—like Romeo and Juliet—and *soldiers*—like the stage versions of Henry V or Richard II.

Parenthetically, we should note a tendency among elders to look critically at the younger generations. This sentiment goes back at least to the time of the Roman playwright Plautus, who reportedly quipped, "Manners are always declining." Far more recently, poet and playwright T. S. Eliot noted, "We can assert with some confidence that our own period is one of decline; that the standards of culture are lower than they were fifty years ago; and that the evidence of this decline are visible in every department of human activity."[1] Fortunately, in the case of our study, harsh judgments that Howard the Elder might be inclined to make are balanced by the more upbeat views of Katie the Younger and Molly the Youngest.

Those, then, are biological or genealogical generations. Once historians, sociologists, and literary critics came on the scene, a new incarnation of generations appeared. Generations came to be associated not merely with those who gave birth to you, or those with whom you shared a dwelling, but also with the kinds of experiences that you shared with peers. We argue here that, in our own time, the digital technologies usher in a new sense of the concept *generation*—one that has implications both for the length of a generation and how its consciousness may be affected. Specifically, the emergence of digital technology in general—and of apps in particular—has produced a unique generation: wrought by technology, fundamentally different in consciousness from its predecessors, and, just possibly, ushering in a series of ever shorter, technologically defined generations.

The feeling of common experiences of members of a generation is brought forth evocatively in Gustav Flaubert's novel of mid-nineteenth-century France, *L'Éducation sentimentale*. On the surface, the novel is about the desires, aspirations, and anxieties of the protagonist Frédéric Moreau, as he attempts to find a career, companionship, romance, love, financial security, and a recognized place in Parisian society. Much of the novel consists of Frédéric hanging around his male peers who are also in search of a calling or place in life, along with more momentary pleasures in games of chance, nighttime chats, and love affairs. They talk about almost everything imaginable—art, music, literature, philosophy, religion, economics, politics—from communism and socialism to monarchical regimes. (For future reference: Note the importance of conversation—in Parisian circles, the cognitive equivalent of breathing!) We learn of their aspirations as they approach adulthood and of their disappointments and regrets as they reach midlife. Flaubert was after even bigger game: "I want to write the moral history of the men of my generation—or, more accurately, the history of their *feelings*. It's a book about love, about passion; but passion such as can exist nowadays—that is to say, inactive."[2]

Flaubert may have been a pioneer in the literary evocation of a generation. In his dramatic treatment of Sturm und Drang (storm and stress), the German literary giant Johann Wolfgang von Goethe was another—but by the early twentieth century, it had become common to describe young people in terms neither of their parents nor of their date of birth but rather of their common experiences. Following many years of relative peace for those not in the professional military, the eruption in Europe of the First World War gave rise to the "generation of 1914," millions of whom died in trench warfare or were forever scarred by their experiences in battle or, less commonly, by their avoidance of combat. Only a few short years thereafter, American writer Gertrude Stein looked at her fellow expatriates in postwar Paris and famously declared, "All of you who served in the war . . . you are a lost generation."[3]

By the time of the Great Depression of the 1930s, membership in a generation was no longer thought of primarily in gendered or geographical terms. Individuals from families that were once comfortable, or at least employed, confronted a new reality in which neither shelter nor employment nor even, at times, food was guaranteed. Genuine recovery for the multitudes did not occur until a decade later and was clearly stimulated by mobilization for the second major war of the century. The national unity spurred by the struggle against Fascism led, in retrospect, to the epithet "The Greatest Generation."

A consequence of the emergence of mass media was the proliferation of generational characterizations both in the United States and abroad. The fifties saw the Silent Generation and the Beat Generation, the sixties saw the rise of hippies, flower children, young radicals, and the stark epithet "The Sixties Generation" . . . and so on, leading to the perhaps deliberately non-revealing appellations Generations X, Y, and Z of recent decades. Indeed, calendrical considerations created pressures for each decade to feature a separate characterization of youth—from the Silent Fifties to the Revolutionary Sixties to the Conservative Seventies.[4]

LONELINESS AND IDENTITY: LITERARY GUIDES TO THE MIDDLE OF THE TWENTIETH CENTURY

For analytical and expository purposes, it is helpful to choose a point in time with which to compare our current app-suffused era. We believe that the best time is the middle of the twentieth century; the best "place" is the America of the middle class. We select this era because it is the last time that one can write about society without explicit reference to computers; the time that Howard—our specimen digital immigrant—grew up; and the publication date of two important literary guides—one from sociology, one from psychology—that have been key in framing our inquiry.

The Lonely Crowd, a sociological study by David Riesman and his colleagues published in 1950, captured this period memorably.[5] On their account, earlier periods in American history were dominated by two forms of national character. The *tradition-directed individual* looked to the examples of those who came in preceding generations for patterns of what to believe and how to behave. We may think in this context of families from the Old World whose patriarchal and matriarchal figures in effect dictated what the younger generation should and should not do. Using the parental generation as a point of departure, the *inner-directed individual* attempted to develop an internal compass that came to govern his or her behavior and belief systems. Prototypical inner-directed individuals cut their ties from home and went to seek fame and fortune in the Wild West, the big city, or the recesses of their own imaginative powers.

The newly emerging *other-directed individual* took cues neither—in tradition-directed form—from those who came before nor—in inner-directed fashion—from a self-constructed value system. Rather, in Riesman's formulation, as models for belief, action, and, above all, consumption, other-directed individuals looked to the examples of their neighbors and to those peers, role models, and certified "experts" about whom they acquired information via the media. A powerful force in bringing about an other-directed mentality were the mass media—radio, television, and perhaps most powerfully Hollywood movies—each of which had come into its own by that time and had become part of the shared consciousness of the nation, if not the world beyond its borders. (We cannot help wondering whether, if Riesman and colleagues were updating their book today, they would introduce a fourth form of character: "app-directed.")

Indeed, the presidency of Dwight Eisenhower (1953–1961) stands out as a time of conformity and at least superficial agreement on what constitutes a viable society, a good society. The era of *The Organization Man, The Wise Men, The Uncommitted,* and *The Power Elite* (to cite just a few of the best-selling book titles chronicling the period) was marked by a relative lack of turbulence; an acceptance of authority from the center (rather than from earlier generations or from one's own internal gyroscope); a proclivity to tend one's own garden without undue immersion in politics; and a perhaps studied avoidance of what we later came to term hot-button issues (race, sex, and, yes, the war between the generations).[6]

As it happens, another equally influential book *Childhood and Society,* by the psychoanalyst Erik Erikson, was also published in 1950.[7] While *The Lonely Crowd* is remembered for the three forms of directedness that have characterized American society over the years, *Childhood and Society* is distinguished for its delineation of eight principal crises confronted by individuals everywhere over the course of their lives. In each case, the life crisis or tension is inevitable; it cannot be bypassed altogether. And should attempts be made to short-circuit the crisis or to end it prematurely, its lack of adequate, settled resolution will haunt individuals for the remainder of their lives.

Of particular interest for our inquiry are the three crises confronted by young people as they emerge from the years of middle childhood to the years of adult maturity. According to Erikson, the first of these "adolescence and beyond" crises surrounds the challenge of identity formation. Beyond childhood, each of us must forge a persona that fits comfortably with our own desires and aspirations; at the same time, the formation of identity cannot be solipsistic—it must also make sense to the surrounding community. It is permissible to have an extended period of identity formation, sometimes characterized by the formidable descriptor "psychosocial moratorium." But if identity is not properly formed and expressed, far less palatable outcomes ensue. One may settle for a mixture of inadequately formulated identities, called "identity diffusion" or "role diffusion" (the burden of the "organization man" in the big corporation or of Arthur Miller's rootless traveling salesman, Willy Loman), or one may end up forging an identity that opposes the major values of the society, called a "negative identity" (the burden of *The Wild Ones* on their motorcycles or of Willy Loman's feckless and rebellious sons). And because an unresolved or inadequately solved crisis affects life downstream, individuals lacking a solid coherent identity have difficulty in forming intimate relations, rearing the next generation, forging new paths, and achieving satisfying closure at the end of life.

Following the resolution—adequate or inadequate—of the "identity crisis," the next challenge is the consolidation of a sense of intimacy: the capacity to have deep, meaningful relations with others, and especially with the significant other, usually one's spouse. In the world described by Erikson, it is crucial to be able to have a multifaceted, abiding relationship with one or a few other individuals. In its absence, one ends up feeling isolated, alone, disconnected. As we'll explore later, experts on digital technologies have speculated that, despite their many electronic connections to one another, many young people today paradoxically have a sense of isolation.

The conflicts of middle life—say, the decades of one's thirties through the decades of the fifties or sixties—are described by Erikson as ones involving generativity versus stagnation. *Generativity* has a literal meaning: the generative individual forms a family and raises the next generation of offspring as well as guiding others for whom one has responsibility. Generativity can also have a broader connotation; rather than simply repeating what has happened in earlier times, the generative individual is able to use his or her knowledge and skills to initiate new thoughts, open up new venues, make a contribution to society, and lead a life that may inspire others. On the downside, for one or another reason, the middle-aged person may be unable to have a family of any sort and may be equally stymied in the deployment of his or her creative and imaginative powers. Like a motionless body of water, such a middle-aged life is stagnant, in permanent "idle." One recalls Biff Loman's lament to his mother, Linda, "I just can't take hold, Mom. I can't take hold of some kind of a life."[8] In our own study, we have focused on those cognitive capacities that enable individuals to think and act in new ways, going beyond and sometimes in contradiction to the paths followed by tradition or by others: we've termed them "imaginative powers."

An aside: Although it occurs before adolescence, the fourth life crisis (industry versus inferiority) may be relevant to our study. The industrious young person masters the various tasks and challenges of the society—in the case of modern society, primarily those challenges posed in school. If one negotiates these well, one should be on the way to a relatively smooth adolescence. One might speculate that the ability to use apps, to master the ensemble of apps, smoothes the way to adolescence—so long as the apps are well understood and used appropriately. But what constitutes appropriate use is not self-evident. Adapting Riesman's terminology to today's society, we believe more and more young people are app-dependent than app-enabled.

MEDIA AT MIDCENTURY: HOWARD'S WORLD

Against the sociological background provided by David Riesman and his colleagues and the psychological landscape sketched by Erik Erikson, what can we say about the media-technology milieu in America in the middle decades of the twentieth century—when Howard grew up? As we have mentioned, the era of the 1940s and 1950s was dominated by the mass media. Radio and movies (first silent films, then talkies, then films in Technicolor) were already part of the cultural landscape; and television was quickly becoming an even more powerful medium, dominating eyes and ears in most households and essentially constituting a monopoly in the hands of three networks—CBS, ABC, and NBC. Metaphorically speaking, the whole country would tune in to comedy shows like *I Love Lucy,* variety shows like *The Ed Sullivan Show,* quiz shows like *The $64,000 Question,* and dramas of the serious *Playhouse 90* or more popular *Gunsmoke* ilk. In a way that is difficult for younger persons to appreciate today, news was presented every evening, in the 1950s for fifteen minutes, thereafter for half an hour; and if you wanted to know what was happening in the world, you would tune in to hear Walter Cronkite (CBS), Howard K. Smith (ABC), or Chet Huntley and David Brinkley (NBC) report the day's events in mellifluous, masculine midwestern tones. Indeed, Cronkite finished each nightly broadcast with the authoritative phrase "That's the way it is"—and if you were not quite sure that you understood the way it was, Cronkite's sober sidekick Eric Sevareid was on hand, ready to explain it to you. For those who preferred to get the news and entertainment from print and still photographs, the media emanating from the Henry Luce publishing empire—*Time, Life, Fortune, Sports Illustrated*— had an informing power and pervasiveness that has not been approached since.

Of course, these media were almost entirely receptive. Except for the very few individuals who created "content," the overwhelming majority of the population participated as consumers. By midcentury, powerful computers were already being built, but they were seen as the province of science and military, as well as—at quadrennial election time—projectors of who would win the major political contests. On a daily, interactive basis, computers were the concern of only a tiny circle of scientists and technologists gathered in American cities like Cambridge, Palo Alto, and Princeton (and in British cities like Manchester, Edinburgh, and Cambridge). Similarly, for every individual who was an active ham radio operator, many thousands of users never thought to assemble or disassemble a radio or television set.

Put briefly, the media offered relatively few choices and very little conflict. People across the land spent much time watching or listening to the same media or reading the same mass publications, typically at the same time. In that sense they readily fashioned a population that was "other-directed"—that shared a sensibility with its geographical and technological neighbors. But at least in Howard's experience, these media of communication did not dominate life in the ways that they have come to in more recent times. There was still time to play, to daydream, to create things on one's own. When children went to camp in the summer, they rarely had contact with parents except by the US mail service, and when they traveled abroad or went away to college, contact was weekly or even less frequently. The concept of a helicopter parent was decades away.

The situation could not be more different from that which obtains today. Howard has taught students intermittently in the 1960s and 1970s and regularly ever since. With every passing decade, it appears to Howard that students look increasingly to their teachers—and more broadly, to their supervisors and their mentors—for the correct way, for what is wanted, for the route to an "A," to approval, to a positive letter of recommendation, smoothing the way to the next step on the ladder of success. There's more. Many students convey the impression that the authority figures know just what they want from their charges; that they could be straightforward and say what is wanted; and that they are being irresponsible, delinquent, unfair, and even unethical in withholding the recipe, the road map. The light-hearted version of this attitude is the all-too-familiar question, "Will this be on the exam?" The nuts-and-bolts version is, "Just tell us what you want and we will give it to you." Even tougher, "If you don't tell us what you want and how to deliver it, we'll get our parents out after you and sue the university—and you."

In our terms, the students are searching for the relevant app. The app exists, the teacher certainly knows it, and fair play entails providing it to the students, as efficiently and straightforwardly as possible. To be sure, given acquiescence on the part of the teacher, the students face a choice. They can use the app in the way that they believe the teacher wants them to use it. But of course the students also have the freedom to use the app flexibly or even to tweak it in new and unexpected directions. The prescient teacher can signal which option he or she prefers.

Of course, the United States has long been a large and diverse country, with millions of young people growing up at any one time. (The same can be said of France and Germany, to cite two other countries mentioned in our narrative.) Almost any generalizations about youth are likely to invite—

and deserve—modifications as well as counterexamples. To contextualize our portrait of youth in the middle of the last century, it should be said at the outset that Riesman, Erikson, and their colleagues are describing middle-class youth—not necessarily youth of wealth, and certainly not extreme inherited wealth, but youth who had access to education and were not caught in an endless cycle (or "culture") of poverty. They were more likely to have been males than females, more likely to have been white than of color, and more likely to have had lofty than middling aspirations, whether or not these were actually achieved. At the same time, we believe that the portrait we've sketched here has reasonably broad applicability, particularly as a comparison to youth growing up a half century later (and, for that matter, youth growing up in earlier generations).

TECHNOLOGIES AND GENERATIONS

So much for the America in which Howard grew up. By the 1960s, in the technological sphere, advances in the media of communication, knowledge creation, and knowledge dispersion were rapid, even dizzying. Led and spurred by Silicon Valley in northern California, echoed in the concentric circles around other large cities in North America, Europe, East Asia, and Israel, the world entered—sometimes invisibly, sometimes ceremonially, sometimes dramatically—the Digital Age. Mainframe computers were followed—and oft-times replaced—by increasingly small and powerful desktop computers; and these in turn, came to be supplanted by laptops, tablets, personal assistants, smartphones, and other handheld devices. Mainframe computers were bulky and clunky. The newer devices had more power and portability, and they operated much more speedily. The hegemony of the major broadcast networks was broken, as cable TV ushered in a proliferation of new channels, many heavily dependent on digital technologies. Perhaps most important, the various digital devices were no longer independent, non-communicating entities. Increasingly, single devices could carry out many functions, and such devices were able to communicate with one another.

Back to our story about the generations, but with an unexpected twist. In mid-twentieth-century America, generations were routinely spoken of in terms of their defining political experiences or powerful cultural forces. Only in recent memory has characterization of a generation taken on a distinctly technological flavor. In his studies of successive waves of college students, Arthur Levine (with colleagues) has discerned a revealing trend. Students in the latter decades of the twentieth century characterized themselves in terms of their common experiences vis-à-vis the Kennedy assassination, the Vietnam War, the Watergate burglary and investigation, the shuttle disaster, the attack on the Twin Towers in September 2001. But once the opening years of the twenty-first century had passed, political events increasingly took a back seat. Instead, young people spoke about the common experiences of their generation in terms of the Internet, the web, handheld devices, and smartphones, along with the social and cultural connections that they enabled—most prominently, the social networking platform Facebook.[9]

Now it could be that we are living in an exceptional time, one particularly—surprisingly, unprecedentedly, perhaps uniquely—inflected by technological innovations. Should this be the case, we can anticipate that future generations may return to self-characterizations in terms of more traditional political, social, and cultural events. But it could also be that young people have shifted sharply—and maybe permanently—from political events as defining; they think of themselves increasingly as part and parcel of the latest, most trendy, most powerful technological devices. Neither Jacques Ellul nor Lewis Mumford would be surprised. We can't know which is the case. We *do* know that this is how individuals born, roughly speaking, in Molly's time—say, from 1990 to 2000—choose to describe themselves to pollsters and researchers in the social sciences.

Indeed, we may be straddling one of those fault lines in history when the definition of a generation needs to be recalibrated. If, in fact, our era is defined in terms of technology, then a generation may be quite brief; indeed, we should think of a generation as that era in which certain technologies rise to the fore and, in particular, when young people—usually the "early adopters"—come to employ particular technologies in a full, natural, seamless, "native" way. If we take this timetable with a generous dollop of seasoning, consider the emergence and widespread use in succession of these various electronic and digital technologies and media:

Telegraph, telephone: late nineteenth, early twentieth
centuries
Radio, movies: 1920s–1940s
Network television: 1950–1960s
Cable television: 1970s onward
Personal computers: 1980s onward
Internet, email, World Wide Web: 1990s onward
Digital consumption (eBay, Amazon): middle 1990s
   onward

The twenty-first century
Web 2.0—blogs, wikis, social networking sites
Multiuser games and other virtual worlds
Texting and instant messaging
Facebook, Twitter, Tumblr, Pinterest, Instagram
Proliferation of apps

Simple calculations, even eyeballing, suggest that a *technological generation* may be much briefer than earlier genealogical, political, economic, or cultural generations. Indeed,
although Katie is only seventeen years older than Molly, her
experience of technology has been radically different. Nor does
this short-lived generation necessarily correlate with calendrical, political, or cultural considerations. Indeed, a single
decade might harbor a number of technological generations,
even as a single powerful technological change like the web
certainly cuts across the decades.

We've even heard some young people restrict the word *technology* to instances of hardware or software that came into

existence during their own conscious lives. If already present when they were very young, the form of technology is just part of the background scenery. On this wry reading, the history of technology is divided into two phases: "What I remember as it erupted on the scene" versus "All previous inventions"!

Another possibility: Going forward, we may need to think of generations operating on a number of quasi-independent timescales. There is the biological generation, defined by child birth; the calendrical generation, defined by decades (or quarter centuries); the political, cultural, or social generation defined by Traditional Big Events; and the technological generation, marked by newly emerging technologies or significantly different kinds of relations to already existing technologies. As we reflect on and attempt to characterize generations, we need to bear in mind these competing definitions, along with the tensions and confluences across them.

To this congeries of "generation talk," we'd add one final consideration. Scholars and observers from a variety of perspectives have converged on the view that, in developed countries, adolescence has recently been extended in length; in the phrase favored by some, there is a new phase of "emerging adulthood."[10] This re-computation of long-standing, well-entrenched life cycle distinctions has been brought out by lengthier educational tracks, a challenging job market, limited family resources, and shrinking safety nets. And so it is far more common today than it was twenty-five years ago for young people in their twenties to live at the parental home, whether or not they can contribute in any way to the family income. It's entirely possible to have persons aged ten, seventeen, twenty-five, forty, and sixty living under the same roof, even though their relations to technology can be dramatically different.

If our analysis is on the mark, we now have a new perspective on the generational issue: invoking the spirit of Marshall McLuhan, we can think of generations in terms of the dominant media and the habits of mind, behavior, presentation of self, and relation to others that they foster—as well as those that they minimize or even expunge.

It is easy and straightforward to speak of the generation of the past decade or so as the "digital" or "web" generation. But in our view, that focuses misleadingly on the technology per se. In invoking the epithet the App Generation, we seek to go beyond the technology, and beyond the media of communication, into the psychology of the users. In the spirit of Jacques Ellul, we aim to capture the cognitive, social, emotional, and even ethical dimensions of what it is like to be a young person today. Living in a world in which there are so many applications at one's fingertips, and so many new ones emerging each month, one is led, perhaps ineluctably, to the following conclusion: What we think, say, do, and dream for ourselves, and how we relate to others, are most perspicuously thought of as apps, whether we are thinking about what to do in the next minute, in the next day, or—in super-app fashion—for the rest of our lives.

While Howard was viewing the Promised Digital Land from afar, Katie, growing up three decades later, saw it much closer—and Molly was thrust right in the middle of it, with little sense of what it was like to live in a time permeated by mass media but innocent of digital hegemony.

KATIE'S AND MOLLY'S UNIVERSES

Katie's youth in the 1980s and early 1990s took place amid the ever-shrinking and increasingly popular personal computer; the rise of cable television, along with the 24/7 news cycle and reality TV shows it spawned; the gradual decline of pay phones and landlines and their replacement by mobile phones; and, most memorably for Katie, the introduction of the World Wide Web, version 1.0.

As a result of growing up on a small island and in a household with a tight budget, Katie's experience of these trends lagged behind many of her American counterparts. The "big three" television networks reigned supreme throughout her childhood and most of her adolescence. Much as Walter Cronkite had done for Howard and his peers, CBS news anchor Dan Rather informed Katie about the *Challenger* crash, the fall of the Berlin Wall, the student protests in Tiananmen Square, and the alarming spread of AIDS in the United States.

Things had started to change by the time the first Gulf War started (and ended) in 1991. Notably, Katie's father and stepmother had gotten cable TV installed in their house. Every Sunday and Wednesday when Katie visited, the three of them watched raptly as Operation Desert Storm unfolded in real time on CNN.

CNN was still the go-to source of round-the-clock news coverage when the Twin Towers fell ten years later. Molly was only five in 2001, but her memories of the event are vivid. Even though she was in a different country and many miles away, the 24/7 news cycle (and her mother's journalistic background and penchant for the news) made it difficult to escape the images and sounds from the tragedy and its aftermath.

Keeping track of—and increasingly contributing to—the news continued to change at warp speed throughout the opening decade of the new millennium. In 2006, Myspace helped students organize a massive, nationwide protest against proposed immigration legislation. Later that year, Saddam Hussein's execution was caught on a mobile phone, and within hours the video was posted on the Internet. The 2008 presidential election was widely dubbed the Facebook Election, the candidates having learned from Howard Dean's successful use of social media to raise awareness and money for his 2004 presidential campaign. And a mere five years after its inception, Twitter was famously used by protesters and journalists during the Arab Spring in 2011.

These digital media contributed to Molly's growing consciousness of the world outside of Bermuda. Compared to Katie's youth, her experiences of this wider world and the events therein have been more vivid, immediate, and interactive.

Like most kids her age, Molly is more likely to use her digital devices to participate and keep track of pop culture than to follow political events as they unfold. Throughout her childhood, reality television has represented a large slice of her pop culture diet. Though it reached a critical mass only in the twenty-first century, reality TV traces its roots at least as far back as the 1992 debut of *The Real World,* MTV's landmark reality series about a group of twenty-somethings living together under one roof. Lacking regular access to cable in the 1990s, Katie never watched the show during her youth, but she and Molly recently came across the first episode of the first season while browsing the television shows on Hulu (Molly watches most of her television on such video-streaming sites). Both sisters were struck by how civil the participants were to each other, as well as by the show's lack of structure and story arc. This episode stands in stark contrast to its modern incarnation and the scores of other reality series that have sprung up since then. Today's shows pivot on high drama, whether it's competing to be the last survivor on a remote island, the last woman standing in a battle to wed an eligible bachelor, or America's next top model, fashion designer, performer, or chef.

The Internet made it possible for Molly and Katie to watch a 1992 episode of *The Real World* on Molly's laptop in 2012. For Molly, that statement is unremarkable because the Internet itself is unremarkable. But for Katie, it still seems a bit magical. After all, she knew nothing of the Internet until 1995, her senior year of high school, when her English teacher took the class on a "field trip" downstairs to the school library and introduced them to the World Wide Web. With great fanfare, he opened up a Netscape Navigator browser and typed in the web address for a site dedicated to Shakespeare's sonnets.

The considerably older teacher who co-taught the course took one look at the small font, clashing colors, and pop-up ads and dismissed the whole thing out of hand, saying, "It'll never last." He pointed out that all of this information was already available in books. He insisted that no one would ever choose to read on a screen instead of in print. He questioned how one could determine the credibility of anything posted on the web.

Needless to say, he was wrong (as wrong as IBM's CEO Thomas Watson, who had been quoted as predicting—either in 1943 or in 1958; sources do not agree—that there would be need for only five mainframe computers in the world!).[11] The Internet has grown from just sixteen million users worldwide in 1995 to well over two billion in 2012. We can do far more online than anyone imagined in 1995. No longer regarded primarily as a content-delivery system, the Internet is highly dynamic and participatory. The problem of screen-reading is largely resolved (thanks in part to e-reader apps like Kindle and Stanza). Credibility issues remain, but with the likes of the *BBC* and the *New York Times* online, plenty of reputable sites do exist.

As Katie was being introduced to the World Wide Web at school, her mother was eight months' pregnant with her second daughter, who would be born in January 1996 without any knowledge of a pre-Internet world. Whereas Katie didn't get her first email account until her first year of college, her first laptop a year after that, and her first cell phone about eight years after that, Molly has trouble remembering any of these firsts. In this way she resembles the dozens of youth whom we studied directly or learned about from our interviews of informed adults. Her struggles around issues of identity, intimacy, and imagination will be played out against a background that could not have been envisioned a half century ago.

We've now provided the promised backdrops for our discoveries. To begin, we furnished a lexicon so that we could examine the contributions of media and technology to behaviors and consciousness in earlier eras. We've gone back to the traditional biological meaning of "generation" and contrasted that to more recent descriptions in terms of consciousness and technology. Then, drawing on major sociological and psychological studies of America in the first half of the twentieth century and with an eye toward the different worlds in which Howard, Katie, and Molly have grown up, we have contrasted the mass media world in the 1950s with the increasingly dominant digital milieu of recent decades.

It's high time for us to look directly at the three aspects of the lives of young people that have been most affected by the digital technology: their sense of identity, their capacity for intimate relations, and their imaginative powers. Later, in conclusion, we'll return to reflect on the effects that the app consciousness may have on other aspects of life today and, more speculatively, on the lives of future generations.

FOUR

Personal Identity in the Age of the App

THE APPS ARRAYED ON a person's smartphone or tablet represent a fingerprint of sorts—only instead of a unique pattern of ridges, it's the combination of interests, habits, and social connections that identify that person. A news app might be sandwiched between a fantasy sports app and a piano keyboard app, revealing multiple facets of one's identity. Because many of these apps provide access to various online communities, each facet allows the owner to find ready communion with similarly oriented people. Though the range of self-expression is great online, it's not unrestricted. For instance, expressions are limited to 140 characters on Twitter, whereas digitally manipulated photos are the coin of the realm on Instagram. The app identity, then, is multifaceted, highly personalized, outward-facing, and constrained by the programming decisions of the app designer.

Just how are youth's identities shaped and expressed in the age of the app? Are they truly different or just superficially so?

We approached these questions in a number of ways, including through extensive interviews and conversations with veteran educators. We found that, as suggested by the app icon itself, the identities of young people are increasingly packaged. That is, they are developed and put forth so that they convey a certain desirable—indeed, determinedly upbeat—image of the person in question. This packaging has the consequence of minimizing a focus on an inner life, on personal conflicts and struggles, on quiet reflection and personal planning; and as the young person approaches maturity, this packaging discourages the taking of risks of any sort. On the more positive side, there is also a broadening of acceptable identities (it's OK to be a geek; it's OK to be gay). Overall, life in an app-suffused society yields not only many small features of a person's identity but also a push toward an overall packaged sense of self—as it were, an omnibus app.

MY TUMBLR, MYSELF

Digital media technologies have given rise to a plethora of new tools and contexts for youth to express and explore their identities—from social network sites, instant messaging platforms, and video-sharing sites, to blogs, vlogs, and virtual worlds. A growing number of youth enter these contexts through an app on their smartphone or tablet. The app interface then becomes an integral part of the way they choose to express themselves online. There are also apps designed specifically for the purpose of encouraging users to play around with their online self-expression. One of Molly's favorites (at least at the time of writing), My Monster Voice, lets users manipulate their voice so that it sounds like one of a variety of preset monster sounds.

In the early days of the Web, scholars probed the many opportunities that the Internet opened up for identity exploration. In her seminal book *Life on the Screen,* published in 1995, MIT scholar Sherry Turkle depicted online spaces as identity playgrounds that give people the freedom to try on identities bearing little resemblance to their offline selves and lacking any repercussions in the physical world.[1] Suddenly, it became possible to alter—with little effort—one's physical appearance, such as gender, eye and hair color, and height and weight, as well as aspects of one's personality, such as sense of humor and level of extraversion. Indeed, if one wanted, it was now possible to become a completely different species!

Turkle focused her initial investigation on the Internet users and online spaces of the mid-nineties, including online chat rooms and multiplayer virtual worlds (then known as MUDs for Multi-User Dungeons). The web has changed considerably over the past fifteen years. Social networking and microblogging sites such as Facebook, Twitter, Pinterest, and Tumblr (each available in app form) have overshadowed the early—and far less populous—online chat rooms like Usenet. Now that a large proportion of the population in developed countries is online, one's friends and followers on these sites are likely to be known offline. Communication on these newer sites is also far more visual, as smartphones and high-speed Internet make it easy to capture and share images and videos.

As a result of these developments, people are more identifiable online, their online lives more interwoven with their offline lives. Indeed, today's young people seldom make a distinction between their online and offline selves. As one of our interview participants, seventeen-year-old Janelle, told Katie: "I am the same person online and offline. So, what you get online is what you get offline, and what you get offline is what you can get online, as well. I'm not two different people."

While there may be consistency between youth's online and offline selves, it's not necessarily a one-to-one correspondence. We gathered considerable evidence that youth take care to present a socially desirable, *polished self* online.[2] Features such as asynchronicity and anonymity (or at least the *feeling* of anonymity) allow young people to craft strategic self-presentations by deciding what information to highlight, downplay, exaggerate, or leave out entirely. Molly has firsthand experience with such carefully crafted online personas. She finds the endless stream of flattering photos, fun with friends, exciting vacations, and accomplishments on her Facebook newsfeed to be exhausting and alienating. "On Facebook, people are more concerned with making it look like they're living rather than actually living."

Supporting the idea that youth present shined-up versions of themselves online, several of our subjects echoed the observation, made by Jenni, a high school senior: "I think everything about you kind of seems, like, glammed up [on

Facebook]. It is kind of, like, the rose-colored glasses, because people aren't going to share about how they got kicked off the soccer team because they were late for every school practice. They are going to share about how their team won; they led their team to victory."

Note that, in seeking glammed-up versions of themselves online, Jenni and her peers don't include every detail about their lives. Such online omissions were common among the youth we interviewed. They consistently told us that their online identity is less complete than their offline identity. As fifteen-year-old Adam noted: "You wouldn't get my whole life story [by looking at my Facebook profile]." To explain this incompleteness, some youth pointed to the constraints of certain platforms and computer-mediated communication in general. A more common explanation, however, related to youth's privacy concerns. Though several youth identified strangers as the people from whom they most wanted privacy, more youth actually identified known others, such as their friends and family.[3] Jenni explained, "On Facebook, there are things that you don't want to tell people; like, you don't want to put up personal things for your random friends to see."

Whatever may have been the case fifteen or twenty years ago, donning wholly different identities online may not be the norm among today's youth. Still, there's certainly ample room for exploration and experimentation. Consider Molly's trajectory of online activities. Where Facebook once represented the central hub of her online activities, she now finds herself spending less time there as her online presence has broadened to sites like Twitter, Tumblr, and Instagram. At one point, she tried her hand at blogging about boarding school life. More recently, she joined Twitter and Tumblr. On Twitter, Molly follows comedians and actors, and she herself tweets, as she puts it, "sarcastic irrelevancies from my life." Like others her age, Molly uses Tumblr mostly to read and reblog others' posts that relate to her particular blend of pop culture interests.[4] These diverse online venues let Molly express and explore different sides of herself and, through that exploration, contemplate who she might become in the future.

Molly's experiences are consistent with what we heard from several youth we interviewed. Some of them told us that, whereas they're soft-spoken and shy in person, they present a more outgoing or extroverted personality online. And of course, multiplayer virtual worlds remain popular among many youth. Brandon, age sixteen, told us about his experiences playing World of Warcraft. As a human warlock, he explained, "I have to assume the character, so it is more of a darker type person, I suppose, as the warlocks traditionally are darker." He said that "it can be pretty fun" to present himself in a different way in the game world than he's used to presenting himself in offline contexts.

TRANSFORMATIONS IN YOUTH'S IDENTITIES

How, if at all, are these new tools and contexts for self-expression influencing the way young people approach the task of identity development? Should we take at face value the consistency that most youth insist exists between their online and offline selves, or would a deeper, more skeptical probe reveal substantive differences? Do apps open up or constrain identity expression? To explore these questions, we examine how certain aspects of youth identity have changed over the past twenty years. For each area of change, we consider the role that our new media landscape has played. And we keep in mind Erik Erikson's original formulation of a healthy identity—a multifaceted but ultimately coherent sense of self that is personally satisfying while at the same time being recognized and affirmed by the surrounding community.[5]

The Packaged Self

Our focus group participants believe that the identities of today's App Generation are more externally oriented than the identities of predigital youth. For the affluent youth, their focus largely rests on presenting a polished, packaged self that will meet the approval of college admissions officers and prospective employers. They appear to regard themselves increasingly as objects that have quantifiable value to others: an SAT score, a GPA, a collection of varsity letters, trophies, community service certifications, or other awards. One religious leader echoed the sentiments of the other participants in his focus group when he said that, for many young people, "Who am I?" means "What am I going to produce?"

Accompanying this sensibility is a calculated effort to maximize one's value in order to achieve academic and professional success. One participant in a focus group said that when youth are asked what their hopes are, they give "pragmatic, achievable answers" situated in the present or near future such as "a good job" or "a good relationship" more often than was the case with youth from earlier generations. During our conversation with the therapists, a participant declared that many of today's young people suffer from a "planning delusion"—a (mistaken) faith that if they make careful, practical plans, they will face no future challenges or obstacles to success.

We've witnessed this firsthand in our freshman reflection seminars at Harvard. Many students come to college with their lives all mapped out—a super-app. "I'll major in government, join the Institute of Politics, intern in DC in the summer, work for Teach for America, then run for state senator in my home district when I'm twenty-eight." Paths to the likes of Goldman Sachs or McKinsey, architectural design or neurosurgery, follow similar trajectories. Put in Eriksonian terms, the students' identities are prematurely foreclosed because they don't allow themselves space to explore alternatives. Not only is this mentality unrealistic (you might flunk organic chemistry, you might flub your interview at Google), but, importantly, it makes those kids who do *not* have their identities all mapped out—who lack the super-app—feel that they are losing out. And that's a reason why they come to the reflection sessions.

The trend toward stepwise progression along a path to success is reflected in the reports of other scholars. Their research points to a generation of youth that is increasingly career-focused and pragmatic, as well as more issue-oriented and less ideological.[6] Today's youth approach their education as "practical credentialists" who complete the tasks necessary to get the diploma they need to secure a desirable job.[7] They are far more focused on "daily life management" than on developing a long-term purpose. Consider that in 1967, 86 percent of college freshmen said that "developing a meaningful philosophy of life" is "very important" or "essential" to them, compared to just 46 percent in 2012.[8]

The pragmatic, careerist focus of today's college students occurs within the context of a broader societal trend toward individualism and away from a more community-minded, institutional orientation. In his landmark book *Bowling Alone,* political scientist Robert Putnam shows that Americans' participation in various civic institutions, such as bowling leagues, labor unions, and church organizations, has declined steadily across cohorts born after World War II.[9] As these community ties loosen, they're replaced by a "moral freedom" that allows individuals to define for themselves the meaning of a virtuous life and doesn't require them to sacrifice their personal needs and desires in the process.[10]

In one study, researchers found evidence of growing individualism in American culture through a survey of tween-focused television shows airing from 1967 to 2007.[11] Their examination of the values depicted in these shows revealed that fame, an individualistic value, was promoted to a greater degree in the recent shows, which included *Hannah Montana* and *American Idol.* In contrast, communitarian values such as

benevolence, tradition, and community feeling were empha-
sized more in shows from earlier decades, such as *The Andy
Griffith Show* and *I Love Lucy*. The researchers also found
that, in comparison with older cohorts, younger cohorts were
more attuned to individualistic values in TV shows.

This form of individualism of the twenty-first century is
quite different from the inner-directedness esteemed by David
Riesman and his colleagues in the middle of the last century.
Indeed, it seems more a product of "other-directedness"—a
concern with what other persons esteem. But there's a clear
difference. The other-directedness of 1950 came from direct
observations of the nearby Jones family or from portrayals in
the mass media. The present variety of other directedness is
more likely to derive from the configuration of the currently
most popular social app, or from scanning the various profiles
available online, or from a combination of such factors.

Individualism goes hand in hand with a focus on the self,
and there's evidence that today's youth are more self-focused
than youth in decades past. Psychologists often use a test
called the Narcissistic Personality Inventory (NPI) to measure
levels of narcissism, or an inflated sense of self. The NPI includes such items as "If I ruled the world it would be a better
place" and "I can live my life any way I want to." In one study,
researchers found that only 19 percent of college students taking the test in the early 1980s scored above 21 (considered a
high score).[12] By the mid- to late 2000s, fully 30 percent of
students scored over 21. A similar trend was found in surveys
of high school seniors.[13] Compared to students surveyed in

the 1970s, high school students in 2006 reported being more satisfied with themselves and had higher scores on self-liking measures. The rise in volunteerism and social entrepreneurship among today's young people seems at odds with these statistics. It's true that the percentage of youth participating in some form of community service has risen notably in recent decades.[14] Although we see this trend as distinctly positive, we're also mindful that, for many young persons, their motivation may stem more from a desire to pad their resume than to give back to society. Seen in this light, the current rise in volunteerism among today's youth may be a product of the packaged self: it's a box to check off as one follows the super-app of life.

Digital (and Predigital) Media and the Packaged Self

To make sense of youth's growing self-focus, certain non-digital societal trends warrant consideration, such as the increasing competition to get into college and secure a good (or any!) job. These trends motivate youth to put forward their best self in order to compete in what is seen (accurately or inaccurately) as an increasingly winner-take-all society. At the same time, the influence of new media technologies on the packaged self is readily apparent. Consider our previous discussion of the glammed-up and rose-colored identities that young people present online. Digital media give youth the time and tools to craft an attractive identity, as well as an audience to view and respond to it.

Facebook and other social network sites emphasize self-presentation by organizing their sites around users' individual profiles. The standard elements of a profile on Facebook—friend list, profile picture, inventories of personal tastes and activities—are used to package the self for public consumption. Presentation and performance are also central on YouTube, where users become the stars of their own video channels. A few of them—including Justin Bieber, a teen heartthrob whom many adolescent girls would like to "marry"—have earned widespread celebrity for their homemade videos, offering others the misleading promise that anyone with a camera and Internet access can achieve similar renown. The discourse of fame surrounding new media technologies like YouTube parallels the growing emphasis on individualistic values that researchers have observed in tween-focused TV shows.[15] Indeed, they point out that many of these shows engage youth across a variety of media platforms, encouraging their participation by holding out the promise that they, too, can become stars like their favorite TV personalities. Perhaps one needs to add to Erikson's ensemble of possible outcomes of the fifth life crisis a new category: "implausible identity."

The educators of low-income youth were particularly concerned about the impact of reality TV on their students. One educator observed that young people increasingly find their role models on MTV rather than in their family or their neighborhood. These television personalities embody a glamorous, self-centered lifestyle that demands little effort or concern for matters beyond their personal and immediate satisfaction.

Several participants pointed to such cultural influences to explain one educator's observation: "Many of our students, even though they aspire to other things, if they could, they would rather be someone in the entertainment industry, or a sports figure." This view is supported by published research indicating that many teens would rather be the personal assistant to a celebrity than to be themselves a prominent executive, author, or researcher.[16] The desire for a celebrity connection is particularly widespread among unpopular kids and kids with low self-esteem.

Apps also prove instructive in contemplating the rise of the packaged self. Individualism and self-focus are evident in the vast marketplace of apps, which gives youth endless opportunity to personalize their digital experience according to their (at least seemingly) distinct combination of interests, habits, and social connections. Just as no two snowflakes are alike, the same could be said (or claimed) of the array of apps on a person's cell phone. Indeed, the app icon itself is worthy of note. One could argue that the icon serves less to signify the purpose of an app and more to represent a particular brand and the lifestyle, values, and general cachet associated with it. In other words, part of an app's appeal lies in its external representation rather than its internal functionality.

Packaging oneself for others involves an element of performance. An app that gained considerable popularity among teens in 2012 illustrates this performative aspect of identity in a digital age. Snapchat lets users take pictures and short videos with their phone (or other mobile device), add text or drawings, and send them to a fellow Snapchat user for a specified length of time (up to ten seconds) before poof! they vanish magically. After considerable cajoling in 2012, Molly convinced Katie to download Snapchat. Something that stood out immediately for Katie was the stagecraft involved in each Snapchat message that Molly sent to her. A typical message might include a "selfie" of Molly making a funny face, overlaid with a wry comment about a thought she just had. Katie concluded that Snapchat exchanges are not so much a conversation between two people, as with standard text messages, but rather a series of mini-performances for an audience of one.

In addition to its carefully crafted, packaged, performative quality, the externalized self also lends itself to measurement and quantification—increasingly an imperative in today's market-driven, big data societies. Sites like Klout and PeerIndex create "influence scores" for users based on the number of followers or friends they have on social networking and microblogging sites (those with high influence scores can even earn rewards and discounts from companies). Examples of other self-tracking tools include Moodscope, to measure, share, and track one's changing moods; 80Bites, to monitor the number of bites of food one takes each day; and Daily Deeds, to keep track of beneficial habits. Timehop and Rewind.me are two apps that function as memory surrogates by mining your social media data to show you what you tweeted or tumbled in years past. A list of more than five hundred such tools can be found on the primary website of the Quantified Self (QS) movement, a large and growing collection of people from all parts of the world who use and create tools for self-tracking. In addition to exchanging ideas online, QS enthusiasts also gather for an annual Quantified Self conference and regular in-person meetings in dozens of cities across the globe.

One psychologist expressed concern about young persons' constant self-projection and self-tracking online, which she says leaves them with little time for private contemplation or identity construction. She worries that, as a result, the prominence of their internal sense of self (in Riesman's term, "inner-directedness") is dwindling, perhaps to the point of nonexistence.

This lament about the lack of time for quiet reflection has become a common theme among academics and the popular press. Researchers have identified a number of benefits that accrue when a brain is at rest (relatively speaking) and focused inward.[17] The downtime appears to play a restorative role, promoting feelings of well-being and, ultimately, helping individuals to focus their attention more effectively when it's needed. Daydreaming, wandering, and wondering have positive facets. Introspection may be particularly important for young people who are actively figuring out who and what they want to be. Without time and space to ponder alternative ways of being in the world—without breaking away from an app-determined life path—young persons risk prematurely foreclosing their identities, making it less likely that they will achieve a fully realized and personally fulfilling sense of self.[18]

Technology was intended to free up time for unstructured

contemplation, but paradoxically it seems to have had the opposite effect.[19] Moments where we once sat alone with our thoughts, either waiting for an appointment in the doctor's office or commuting to work on the train or walking the dog, have now been replaced by virtually compulsory listening to music, text messaging, or playing games on our digital devices. Far too often, we do these things all at once. We tweet about events as we experience them, reply to one person's email while talking to another, and, particularly popular among youth, juggle multiple text and instant message conversations simultaneously. Epitomizing the purpose of the app, we're more focused on *doing* than on *being*. One psychologist we interviewed commented that because young people are inclined toward constant virtual connection with others, they don't allow themselves the time and space to figure out their thoughts and desires; they are, consequently, "rendered insecure" by lack of self-knowledge.

Several participants echoed another hot topic of debate among the digerati and mainstream media commentators when they questioned explicitly whether young people's digital lives are making them narcissistic. Katie remembers wondering the same thing when Molly first joined Facebook. Using her cell phone or Photo Booth on her MacBook, Molly seemed to be forever posing for an endless stream of self-portraits, or "selfies," which she then posted to Facebook (or at least the most flattering shots). In this vein one educator we interviewed commented: "Facebook and texting, though, it's constant validation. As soon as somebody buzzes you on your phone it's like 'somebody is paying attention to me.' Facebook, 'oh, I got fifty likes on that stupid picture I put up there, I guess people are paying attention to me.' I mean it's so narcissistic, and I'm not saying that [kids] were less narcissistic [before the Internet], there are just more ways to be validated now with that."

The question of the Internet's impact on self-focus has also become a popular focus among social scientists, who've generally observed a positive connection between narcissism and online behavior.[20] For instance, one study found that people with high narcissism scores were more likely to post self-promoting content and engage in high levels of social activity on Facebook.[21] Another found that college students with high narcissism scores were more likely to tweet about themselves.[22] The authors of this study caution that while youth's online behavior may appear narcissistic to an outsider's eye, it's important to keep in mind that their primary motivation for going online may well be not to promote themselves but rather to maintain and nurture their social ties. (We'll examine the social dimension of youth's online lives in the next chapter.) Still, it's worth noting that about 30 to 40 percent of ordinary conversation consists of people talking about themselves, whereas around 80 percent of social media updates are self-focused.[23] Also important is the fact that we can't determine in which direction the arrow of causality points. Does Internet use cause narcissism, or do narcissistic people use the Internet in distinctive ways?

"SCARED TO DEATH"

Given the self-focus of narcissists, one might assume that they're self-assured and unaffected by the goings-on of others. This turns out not to be the case. As Sherry Turkle explains in her book *Alone Together,* "In the psychoanalytic tradition, one speaks about narcissism not to indicate people who love themselves, but a personality so fragile that it needs constant support."[24] Instead of self-assuredness, then, narcissists tend more toward a fragile self that needs propping up by external reassurances. Jean Twenge's research bears this out. Along with rising levels of narcissism among youth, she finds increasing moodiness, restlessness, worry, sadness, and feelings of isolation. In sharp contrast to Riesman's inner-directed persons, today's young people are also more likely to feel that their lives are controlled by external social forces rather than growing out of an internal locus of control. Consistent with Twenge's findings, researchers at the University of California at Los Angeles found that the percentage of first-year college students who said that they frequently felt "overwhelmed by all I had to do" during their senior year of high school increased from 18 percent in 1985 to 30 percent in 2012.[25]

Several of our participants identified a similar incongruity between youth's external polish and their internal insecurities. The camp directors we interviewed told us that campers today demonstrate more self-confidence in what they *say* they can do but are less willing to *test* their abilities through action. They attributed this shift to youth's growing distaste for taking any tangible risk that could end in failure—failure that once might have been witnessed by a few peers and then forgotten but today might become part of one's permanent digital footprint.

The themes of growing anxiety and aversion to risk surfaced in other focus groups. One therapist reflected that youth seem reluctant to engage in certain endeavors for fear of feeling anxious or depressed if they don't go as planned. Indeed, many participants agreed that young people's identities are defined by insecurity and disequilibrium. The religious leaders remarked that youth today are generally more fearful about their future. "Even the most confident Harvard grad . . . ," shared one participant, "is . . . scared to death." The therapists observed that, to cope with this fear, many young persons display a notable lack of affect and an apparent goal to "feel nothing." Citing a word all too familiar to parents of today's teenagers, one participant called today's youth the "'whatever' generation."

This anxiety and desire to "feel nothing" may explain why alcohol abuse, including binge-drinking and driving while intoxicated, has increased dramatically among college students in recent years.[26] It is not uncommon for students at selective colleges to drink steadily from Thursday through Sunday evenings. Possibly reflecting analogous pressures, the educators who work in low-income neighborhoods have observed a growing sense of hopelessness among youth stemming from the increased violence in their neighborhoods and a dramatic decrease in job and advancement opportunities. One educator reflected, "Kids used to get into a fight and a kid might get a nose broken. The difference today is, a kid might shoot a kid and they're dead." Other participants agreed that being surrounded by such violence leads at least some of these youth to take greater risks, believing they have nothing to lose since they're unlikely to make it past their teen years.

Nondigital Sources of Anxiety

The anxieties of today's youth are reflected in large-scale data tracking young people's lifestyle choices. According to a 2012 *New York Times* op-ed, compared to their counterparts from the 1980s, today's young adults are considerably less likely to move to another state.[27] Members of this age cohort were also twice as likely to live at home in 2008 compared to 1980.[28] Today's youth even appear to be reluctant to leave the house for a drive. Fully 80 percent of eighteen-year-olds held a driver's license in the early 1980s. By 2010, less than two-thirds (61 percent) of eighteen-year-olds had earned their driver's license.[29]

As these trends suggest, there's a large, apparently nondigital component to youth's growing anxieties. Several participants reflected on the challenging economic landscape that today's young people face. Indeed, researchers have documented that youth who grow up during a recession are less likely to leave home, take risks with their investments, or start their own company. Young people caught in such circumstances are also more likely to believe that luck, rather than individual effort, plays the biggest role in a person's success.[30]

Our participants also identified the increased focus in education on standardized testing and accountability as a major cause of young people's growing passivity and aversion to risk-taking. Federal initiatives in the United States such as No Child Left Behind and Race to the Top tie government funding to students' test scores, making it necessary for schools to structure the school day around efforts to improve their students' test performance. (The educational testing industry is not unaffected by the app atmosphere of ranking, counting, and prepackaged curricula—we regard this as a vicious rather than a virtuous circle, a triumph of behaviorist approaches over constructivist ones.) This environment discourages risk-taking by placing top priority on filling in the correct answer on a multiple choice test. It's also likely to breed anxiety when students' failure to pass the test may not only diminish their chances for admission to college but also get a teacher fired or even a school closed.

Our focus group participants didn't let parents off the hook, either. They observed that today's parents demonstrate a passionate desire to shield their children from experiencing any sort of unhappiness or hardship. The therapists we interviewed observed that this emphasis on happiness seems to leave young people unable to cope with the emotional complexity of life. One psychiatrist put it this way: he is "an advocate[, not] of unhappiness, but [of] the ability to tolerate unhappiness."[31]

In lower-income families, the desire to shield one's children often takes the form of parents working hard so that their children don't have to. One educator reported that on the day after a major snowstorm, she asked her students to raise their hands if they had shoveled their families' driveways and sidewalks; none raised their hands. She said that according to her students, they spent the day indoors while their parents shoveled snow. In wealthier families (where snow-clearing is outsourced to a local vendor), the desire to shield one's children from disappointment often takes the form of parents micromanaging their children's lives so that they might avoid mistakes and failure. Our participants worried that, through these well-intentioned actions, parents are unwittingly promoting passivity among their children and preventing them from developing a secure sense of autonomy and from taking unsanctioned but reasonable risks.

Hiding behind the Screen

Although changes in the economy, education, and parenting are no doubt important contributors to youth's aversion to risk-taking, it's instructive to consider the role that our digital media environment may play. Not just in the United States—indeed, in fifteen countries!—researchers found that the proportion of youth who are online is inversely related to the proportion of youth with a driver's license.[32] As with the research connecting narcissism and Internet use, this study can't say whether time online causes or is the result of a postponed driver's license. We note, though, that if you spend lots of time poring over Facebook, there's less time—and perhaps less need—to take to the road.

Facebook may keep young people from taking risks outside the house, but what about risk-taking online? Since children started using the Internet in large numbers, there's been considerable attention paid to the harm that might befall them at the hands of online predators. As it turns out, young people are far more likely to be harmed by someone they know offline than an online stranger.[33] Still, many adults interpret youth's online sharing of personal information and interactions with strangers as risk-taking behavior.

Young people are more careful about their online actions than some adults might think. Considerable empirical evidence indicates that youth are both aware of and care about privacy risks online. A 2010 survey of thirteen-to-seventeen-year-olds living in the United States found that 88 percent of teens said they worry about the consequences of posting their contact information online.[34] In our interviews with middle school students, we found that participants' privacy concerns led them to employ a variety of strategies to protect their privacy online, such as using the privacy controls on social networking sites and choosing to omit personal information like their home address and telephone number.[35] In fact, other researchers have found evidence that young people's privacy-protecting behaviors on social networking sites have increased over time such that they are more likely than older adults to engage in these cautionary behaviors.[36]

There's an interesting paradox here. Using privacy controls may give youth the (misleading) impression that it's safe to reveal what lies beneath the external polish they present to adults offline. For example, there's evidence that the rise of excessive alcohol consumption among young people is reflected—and touted—online. In one study, researchers found that over half of the college students in their sample selected an alcohol-related image for their Facebook profile. Overall, the results suggest that many college-age youth find it desirable to express an alcohol-related identity on Facebook.[37]

Snapchat—the app that lets you send self-destructing images to others—represents another example of the false sense of security that youth may feel when interacting with their peers online. Anecdotal evidence suggests that many teens are using Snapchat to send revealing pictures and video clips of themselves.[38] But even Katie and Howard—who are far from being tech wizards—immediately imagined a scenario where the recipient of a sext might use a second device, be it a camera, phone, or tablet, to take a picture of an image sent via Snapchat, rendering it decidedly more permanent than the sender intended. As it turns out, people are already doing this. In early 2013, two teenage girls in New Jersey used Snapchat to send nude photos of themselves to a male classmate, who promptly took screenshots of the photos and posted them on (the very public) photo-sharing site Instagram.[39]

Snapchat and sexting aside, apps and the app mentality in many ways support and reinforce youth's general shift toward risk aversion. There exist a host of apps to remove many of the risks that have represented until now a standard part of daily experience. Messaging apps remove risks associated with interpersonal communication by doing away with the discomfort one might feel when confronting someone face to face. Information apps take away the risk of giving an incorrect answer, whereas location apps eliminate the risk of getting lost in an unfamiliar place. It strikes Katie and Howard as a remarkable fact that Molly has never had the experience of being lost. Each of us can recall instances from our youth when we didn't know where we were and didn't have immediate access to a parent to guide us to familiar territory. Though scary, these experiences stand out in our memories because they tested our resiliency and gave us a sense of autonomy. Such experiences are foreign to Molly. With her map app and ability to call her parents at any time, she can always be sure of where she is and how to get to her next location . . . unless she loses her cell phone!

Let's return for a moment to our participants' observation that today's parents increasingly desire to protect their children from any stress or failure. It's easy enough to see how technology feeds into this desire. Participants said that it's not uncommon for today's college students to text or call their parents multiple times a day. One participant observed, "Students at college are not on their own anymore, with parents seeing their grades and maybe their bank overdrafts. . . . I guess the term for it is 'helicopter parents.'"

As mentioned in our opening discussion, we heard similar concerns from the camp directors we interviewed. Camp has traditionally been seen as an early step toward autonomy, when children first leave home for an extended period. This step is becoming harder for youth to take in today's media-saturated world that allows parents and children to stay connected even when they're physically separated. It's worth recalling the testimony of the camp director who told us that it's all too common for parents to send their children to camp with two cell phones—one to surrender publicly in apparent compliance with the camp's no-tech policy, and one to reserve for surreptitious texts and calls home. As this anecdote suggests, technology only facilitates this constant connection; it doesn't instigate it.

Other researchers have found similar evidence for technology's facilitating role in maintaining high levels of contact between youth and their parents.[40] A series of studies involving undergraduates, recent graduates, and parents revealed that college students were in contact with their parents an average of 13.4 times per week.[41] Exchanges involved parents providing direction on classwork, as well as a "best friend" phenomenon in which children shared their daily goings-on and moment-to-moment feelings with their parents (usually their mothers). Howard cannot conceive of an analogous situation fifty years ago.

Turkle uses the metaphor of a tether to suggest that youth's constant connections to their digital devices and the people accessible through them weaken their ability to develop an autonomous sense of self. These technologies encourage youth to look outside themselves for reassurance, in matters both mundane and existential. Indeed, their thoughts and feelings don't seem real until confirmed by others. This argument is supported by empirical evidence showing that college students who use their digital devices to maintain frequent contact with their parents tend to be less autonomous.[42] In the spirit of Turkle, some scholars have invoked the concept of *psychasthenia* in an effort to explain how people's online presence can weaken their sense of self to the point of full renunciation.[43]

JUST ABOUT "ANYTHING GOES"

Our participants observed and celebrated the fact that today's youth enjoy greater freedom to adopt and rejoice in identities that were either unknown or scorned in decades past. They've become more accepting of those who are different from themselves. These youth are less likely to alienate their peers who vary from the social norm—the "geeky" kids— and they're more accepting of non-heterosexual peers. Racial conflict, while certainly still persistent in some contexts, has largely diminished. One teacher noted that the occurrence of interracial prom dates is now so common at her school as to be unremarkable. This state of affairs contrasts starkly with her early career, when interracial dating was practically unheard of and she was mandated by the school to take attendance based on race! These observations align with Arthur Levine and Diane Dean's research in colleges across the United States;

they find that today's students display greater comfort with racial, ethnic, and diverse gender roles. Similarly, researchers at the University of California at Los Angeles find that fully 75 percent of first-year college students surveyed in 2012 said they support same-sex marriage, an extraordinary rise from 51 percent in 1997, when the question was first asked.[44]

According to the religious leaders we interviewed, the broadening of identities open to youth has had a notable effect on their spiritual lives. They suggested that young people are less likely to embrace membership in a given religious community unquestioningly, particularly the communities in which they were raised. Rather, their sympathies are distributed across multiple interest groups. "Young people . . . feel responsible to the city, the world, South America," reflected one rabbi, discussing the challenge of negotiating multiple identities in today's global, interconnected world. "There's no strong sense of connection to [the Jewish] people. . . . Do they understand that there's a Jewish community?" In the event that youth do try to engage with familial traditions, their spiritual engagement may be based on limited or no faith-specific information. A minister noted that the young Christians with whom he engages are "uninformed about their place in the cosmos" (though he did add that they are also sheepish about their ignorance).

Other researchers have observed a similar pattern. Students are aware of and interested in global perspectives, but most lack an understanding of global issues and are weak in cultural knowledge. One study conducted in the mid-2000s revealed high percentages of young people reporting a lack of even name recognition of many public figures well known at the time, such as then Secretary of the Treasury Henry Paulsen and Chinese leader Hu Jintao.[45] They were far more likely to recognize entertainment figures like the performers Miley Cyrus and Pink. According to the study's authors, today's young people "talk global, but act local." Rather than trying to use the language of the country being visited, it's much easier just to activate a translation app.

The religious leaders suggested a distinction between youth's curiosity about different perspectives, experiences, and practices, on the one hand, and the focused, sustained attention that's required for a deeper understanding, on the other. Indeed, one educator believes that the apparent increase in racial harmony among students comes with a certain cost: "In some ways, it's a good thing that [kids today] have a lot more friends from different backgrounds; they're more accepting, they're more understanding. But I don't know that there's as much of an understanding that a lot of this stuff [racial discrimination] still exists and there's still a tie between racism and economic disparity." Apparently, greater *acceptance* of diverse cultures, lifestyles, and perspectives has not always been accompanied by a greater *understanding* of the source and implications of these differences. We'll explore this lack of understanding and its implications further when we consider the nature of intimacy in today's youth.

DIGITAL EXPOSURE

By connecting us to people and places beyond our immediate geographic setting, technology has played a central role in creating today's globalized world. Apps serve as portals to this world. Whether you prefer to read, listen, or watch, news apps can bring you the latest events from every corner of the globe, while social networking and microblogging apps can bring you the viewpoints of the people living there. An important word here is "can"—the opportunity to broaden our perspectives doesn't necessarily translate into action. Consider that, at the time of this writing, seven of the ten most downloaded free apps on iTunes were games. Moreover, there's evidence from several quarters that people are more likely to visit sites that reinforce rather than challenge their beliefs.[46]

The educators of low-income youth reflected on how digital media have altered their students' awareness of and relationship to the wider world. In many respects this change is positive, as it broadens students' awareness of experiences and opportunities beyond their immediate, circumscribed environments. But there is also a downside to such exposure. One educator noted that because of the Internet and other media, "now kids know they're poor," because they're exposed constantly to privileged lives that aren't theirs. This state of affairs can reinforce a sense of hopelessness even as it may raise some young people's goals and expectations of what they might achieve.

Still, many people—including youth—are optimistic about the Internet's power to expand our horizons and enrich our lives. In his book *Here Comes Everybody,* Clay Shirky suggests that the bowling leagues, lodges and rotary clubs of the fifties and sixties have not simply vanished; rather they have been replaced by a far greater number of online communities representing a wider range of interests.[47] No matter how obscure one's interest, it can find expression and validation online, whether it be down the street or halfway around the globe.[48] For young people, this access to "digital alter-egos" means that their identities as fan girls, gamers, chess players, or knitters don't have to be set aside to fit into a narrow peer culture.[49]

The girls who participated in our study of teen bloggers described the value they place on being able to express their more marginalized identities online. One college freshman, Samantha, used LiveJournal, an online journaling community with a strong fan culture, to participate in fan communities surrounding the *Harry Potter* series and her favorite TV shows, including *The Office* and *Roswell.* She observed, "I can definitely be more of a deranged fan girl in my Live-Journal than I can in real life. . . . I don't have to, like, kind of censor myself. It's not even really censoring myself in real life; it's kind of recognizing that people aren't that interested and kind of dropping off before I freak them out." At least for those who have the skills and the inclination to describe their more personal thoughts, apps can be enabling rather than restricting.

TAKING STOCK

The app metaphor helps us to see how the changes we've explored in youth's identities—increasingly externalized, packaged selves; a growing anxiety and aversion to risk-taking; and a broadening of acceptable identities—are each products of our time. As portals to the world, apps can broaden young people's awareness of and access to experiences and identities beyond their immediate environment.

Whether youth take advantage of these opportunities remains an open question. As visual icons that are selected to personalize one's phone, apps reflect young people's emphasis on external appearances and individualism. Apps also function as safety nets, removing many of the daily risks we previously took for granted, such as confronting a person's unfiltered reaction to a sensitive topic or finding one's way in an unfamiliar locale. These connections between the identities of today's youth and the qualities of apps illustrate our central argument. New media technologies can open up new opportunities for self-expression. But yoking one's identity too closely to certain characteristics of these technologies—and lacking the time, opportunity, or inclination to explore life and lives offline—may result in an impoverished sense of self.

FIVE

Apps and Intimate Relationships

REMEMBER THAT FAMOUS TAG line, "Reach out and touch someone?" AT&T first used it for an ad campaign in the early 1980s to convey the power of telephones to bring people together across geographic boundaries (the corporation was trying to sell long-distance calling at the time—a service that's become increasingly difficult to promote in the age of Skype and other voice-over IP services). By giving us an array of tools, formats, and platforms to connect to others, apps have transformed what it means to reach out and touch someone. Whether exchanging a private joke with one friend through Snapchat or WhatsApp Messenger or sharing a memorable experience with eight hundred friends on Facebook or Tumblr, connecting with others has never been easier—or more constant. Whether, and in which ways, these developments are good or bad for the quality of our interpersonal relationships constitutes the focus of this chapter.

Just how have our deeply rooted, long-term connections to

other people been affected by the unprecedented connectivity afforded by new media technologies? Our investigations suggest that this connectivity certainly has its value—helping friends and family keep in touch when separated by geography; providing opportunities for young people with similar interests to find and interact with one another; and making it easier for some youth to disclose their personal feelings to others.[1] And yet there may be a dark side to mediated communication, as we'll see when we consider the negative consequences of conducting relationships at arm's length, round-the-clock, and simultaneously, and only with those who reinforce one's worldview. Ultimately, we find that the quality of our relationships in this app era depends on whether we use our apps to bypass the discomforts of relating to others or as sometimes risky entry points to the forging of sustained, meaningful interactions.

TALKING WITH TECHNOLOGY

Today's youth communicate in fundamentally different ways from their predigital counterparts. With their ability to transcend geographic and temporal barriers, Internet-enabled cell phones, tablets, and laptops—each with their arsenal of apps for all occasions—have altered what can be said, where, and to whom. Perhaps the most notable change is the constancy and immediacy of communication made possible by mobile technology. As of 2013, the Pew Research Center reports that

78 percent of all adolescents in the United States own a cell phone.[2] This statistic means that for nearly four out of every five American teens, their family members and friends are never more than a text message (or tweet or Snapchat) away. The data on teens' text messaging behaviors suggest that teens take full advantage of this ability to engage in frequent, on-the-run communication. Sixty-three percent of teens say they text every day with people in their lives, and the typical teen sends about sixty text messages per day (among older girls, that number jumps to a hundred).[3] And now, with the widespread use of app-filled smartphones, the range of operations that teens can perform on the go has extended far beyond phone calls and texting.

What are teens saying through their apps, and to whom? As it turns out, a considerable portion of teens' computer-mediated communication is dedicated to making (and sometimes breaking) on-the-fly arrangements to meet up with their friends in person. In one of our studies, we asked teens what they would miss most about not having a cell phone.[4] Sixteen-year-old Justin answered, "Just being able to make plans on the go, and stuff, because me and my friends, we don't really plan things. We just go out." The app mentality supports the belief that just as information, goods, and services are always and immediately accessible, so too are people. Scholars in the mobile communication field have dubbed such in-the-moment planning "microcoordination" and observe that it can slide into "hypercoordination" when teens start to feel left out of their social circles if separated from their mobile devices for any period of time.[5]

Not all mediated communications serve a logistical purpose; many function as "virtual taps on the shoulder," establishing and maintaining a sense of connection among friends who are physically separated.[6] When asked what he and his friends text each other about, one of our study participants, fourteen-year-old Aaron, explained, "Just like, 'How was school? How's life? What you been up to?' because I think texting is, like, an easy way to keep in contact." Aaron noted that such conversations sometimes last throughout the day. There may be breaks while one or both friends go to class or eat dinner, but before long they return to their cell phone screens to resume their conversation. For seventeen-year-old Jenni, texting is a way to fill time when there is nothing else to do: "[Texting] is kind of just like touching base, and just kind of like when I am bored, and I am like, hmm, what can I do? Meghan is always around. I will talk to her."

Though most common among friends and romantic partners, microcoordination and virtual taps on the shoulder have become standard among families as well. Cell phones allow family members to make plans and coordinate their schedules in a more fluid, impromptu way than was feasible in years past. If a teen child decides in the middle of the day that she wants to go to a friend's house after school rather than return home, it's easy to call or text a parent for permission. And, as discussed in the previous chapter, parents can—and do—use cell phones and Facebook to check in with their college-age children and stay plugged into the daily rhythm of their lives.[7]

Accessibility isn't the only new and noteworthy quality associated with today's communication technologies. Social networking sites have transformed many social interactions into considerably more public affairs than they would have been in predigital times. In addition to phone calls and in-person communiqués, the Facebook wall offers a new—and very public—means to plan and chronicle social events and shared experiences. Who's invited to a party and who's not becomes a matter of public record, as do all activities—however cringeworthy—that are captured and uploaded by a cell phone camera. The beginnings and endings of relationships are similarly documented in a far more public manner than in years past.[8]

While Facebook communications are more public than the typical face-to-face conversation, other computer-mediated communications may feel more private to young people. Texting and instant messaging typically involve just two conversation partners, making them more intimate than wall-to-wall exchanges on Facebook. For some youth, these typed exchanges may feel even more intimate than face-to-face conversations. The fact that conversation partners look into a screen instead of each other's eyes, coupled with the fact that—more often than not—they don't occupy the same physical space may make it feel less risky and uncomfortable to share personal feelings with another person.[9] One of the teens we interviewed, fifteen-year-old Christina, explained why she prefers text-based forms of communication: "I am not a good person with feelings, and, like, I am not that good with saying my feelings face to face sometimes because I don't like people to see how I think and what I am feeling." Other research studies have documented a similar preference among youth for text-based forms of self-disclosure.[10]

In many ways, today's social interactions bear markings of the app. Apps exist to maximize convenience, speed, and efficiency. When you want something, it's there for your immediate use. When you're done with it, switch it off (provided you disable the push notifications that will otherwise pop up unsolicited, alerting you to new content). If you tire of an app's wares, delete it. Apps are under our control (though our increasing dependency on them places us in danger of coming under their control); they're available around the clock and seemingly risk free. Much the same could be said of the way today's youth communicate through digital media technologies.

DIFFERENT, YES; BETTER? NOT SO SURE

Young people's social interactions may look quite different today than they did twenty years ago. What is less clear is whether this change in how relationships are conducted has translated into a change in the quality of these relationships. Are the social networks of today's youth larger or smaller, deeper or shallower, than predigital youth? Are their interpersonal relationships more or less authentic, supportive, and fulfilling? As we contemplate these questions, Erik Erikson, introduced in chapter 3, is ever in our mind. The psychoanalyst's model of human development has as the central task of young adulthood the formation of deep, long-term relationships with others; in their absence, feelings of isolation and disconnection supervene. In the latter case it becomes more difficult to negotiate subsequent challenges in life, such as raising a family and launching a successful work life.

AMERICANS' GROWING ISOLATION

To compare the size of Americans' core discussion networks, scholars and journalists invariably point to a sociological study that used data from the General Social Survey (GSS)—an annual survey of Americans' lifestyle choices, values, and beliefs—in 1985 and 2004.[11] The researchers were interested in whether the number of close interpersonal relationships—often called strong ties—enjoyed by Americans had grown or shrunk over this twenty-year period. They looked at responses to the following question: "From time to time, most people discuss *important matters* with other people. Looking back over the last six months—who are the *people* with whom you discussed matters important to you?" In addition to changes in the size of core discussion networks, the researchers were also interested in whether the makeup of these networks had shifted in the last twenty years. Fortunately, the GSS also asks respondents about the demographic characteristics and the nature of their connection (for example, Spouse, Parent, Sibling, Co-Worker) to each discussion partner.

The results are dramatic. In 1985, the average number of discussion partners reported by Americans was 2.94. By 2004, this number had fallen to 2.08, a decline of nearly one person (or, otherwise put, a shrinkage of one third in the ambit of one's discussion circle). The researchers also found that the makeup of core discussion networks had shifted from non-family ties—the sort formed in neighborhood and community contexts—to family-based relationships, especially spouses. Moreover, the number of people reporting that they talk to *no one* about matters they consider important to them rose from 10 percent in 1985 to 25 percent in 2004.

At the same time, there's been a parallel trend toward decreasing trust in others. Since the last quarter of the twentieth century, Americans have become less and less trusting of their fellow citizens and democratic institutions.[12] In 1972, 46 percent of respondents to the GSS agreed with the statement that "most people can be trusted." By 2008, a mere 33 percent of respondents agreed. This dramatic downturn in trusting dispositions has important implications for intimacy and social isolation. If you don't trust that people are playing by the rules, you're less likely to open yourself up to them and become close.

The findings from our analysis of high school students' creative writing and artwork over the last twenty years suggest that youth have also been affected by the trend toward social isolation. Art that features isolation or solitude imagery increased from 15 percent of the artwork from the early 1990s to 25 percent of artwork from the late 2000s. With respect to our creative writing analysis, peers are more frequently present in the recent stories but are also more frequently mentioned in relation to a character's isolation. Specifically, 76 percent of the stories from the early 1990s include no mention or only minimal mention of peers, whereas peers are featured prominently in the majority (60 percent) of the later stories. However, in approximately one third of these later stories, the mention of peers relates directly to a character's isolation, such as a character fantasizing about having friends to play with in the mud or a character with a notable lack of friends (even though the character may be around "peers").

ALWAYS CONNECTED, BUT NOT ALWAYS CONNECTING

The social isolation trends detected in the General Social Survey have been highlighted in books like *Alone Together* and *The Lonely American,* as well as in feature magazine articles like "Is Facebook Making Us Lonely?" and "Are Social Networks Messing with Your Head?" These various texts make the case that Americans have become increasingly lonely and socially isolated.[13] As some of these titles suggest, the blame for this disturbing trend is placed largely at the feet of new media technologies like cell phones, Facebook, Twitter, and email. Indeed, some researchers have found that young people who spend more time with their digital devices experience less social success.[14]

The connection between social isolation and social media isn't obvious. Indeed, it sounds counterintuitive. How can it be that technology designed to connect people may actually be making them feel *less* connected? To puzzle through this apparent paradox, let's consider Molly's experiences with Facebook. During her junior year in high school, Molly decided to deactivate her Facebook profile. She had come to resent feeling pressure to keep abreast of the constant activity of her peers there. "Having Facebook, you just feel like you have to go on it or else you're going to miss something, or if someone writes on your wall you don't want to wait two or three days to reply. So it just feels like you constantly have to be going on." Visiting Facebook also made Molly feel "out of it" when she saw classmates tagging and commenting on each other's "muploads" (pictures taken and uploaded via mobile phone). The pictures and the flurry of comments attached to them painted a picture of a tight-knit group of friends who seemed to be having a lot more fun than Molly night and day. The contrast between this external parade of happiness and the ups and downs of her internal psychic life left her feeling as though she somehow didn't measure up.

Molly's not alone in this feeling. In one study involving college students, researchers identified a connection between respondents' Facebook use and their perceptions of other people's relative happiness.[15] Students who had used Facebook for a longer time and those who spent more time each week on the site tended to agree more that others were happier. In addition, students who included more people whom they didn't personally know as their Facebook "friends" agreed more that others had better lives. So, one possibility is that social media platforms like Facebook make us feel lonely because they create the impression that our "friends" are hanging out with a greater number of more exciting people and having more fun than we are. We've also heard from young people that they spend hours looking at the achievements of peers whom they know only through Facebook and that this voyeuristic activity makes them feel both competitive and inadequate.

Sherry Turkle offers another explanation. Though apps allow us to perform a multitude of operations, they may not be well suited to support the kind of deep connection that sustains and nourishes relationships. By necessity, 140-character messages, the maximum on Twitter, must be stripped down to their essentials (much as apps streamline their content in order to maximize efficiency and speed). To be sure, one can infuse a short message with considerable wit and innuendo (that's the perennial appeal of haiku), but it's next to impossible for discussion partners to communicate and respond to each other's complex feelings in this way.

Turkle also notes that we may deliberately avoid deep communications through text, aware as we are of the fleeting nature of our tweets, texts, and, in the extreme, self-destructing Snapchats, as well as our suspicion that the people on the other end may not be giving us their full attention. One of our study participants, seventeen-year-old William, expressed a similar sentiment when he observed: "People usually, the first thing they are not doing is talking to you [through instant messenger]. They won't be actively talking to you. They will be browsing something on the web and every time they have, like, five minutes, they will just quickly write something to you. So, when you are talking to a person face to face, because you are having an active exchange of information, it makes what you say more meaningful."

This sense of the superficiality of many online interactions was expressed eloquently in a *Huffington Post* blog post written by fourteen-year-old Odelia Kaly. Kaly had deleted her Facebook account a few months before writing her blog post, because, like Molly, she had come to experience Facebook as an alienating space. It made her feel depressed to see pictures of people who looked like they were always having a terrific time, a much better time than she was having. But it wasn't just that she felt inadequate when comparing her experiences with this onslaught of staged happiness and success. For Kaly, the quality of interpersonal communication on Facebook—the stream of obligatory (and perfunctory) happy birthday wishes and wall post "likes"—felt unsatisfying and inauthentic. Though deleting her Facebook account made her feel isolated from her larger peer group, it also gave her perspective on her friendships within that group. She observed, "If deleting my account taught me anything, it was how to weed out my real friends from the fake ones."[16]

An important quality of deep relationship is the vulnerability required from those involved. It's uncomfortable to confront another person directly with one's thoughts and emotions. But taking that emotional risk is what brings us closer to others. We share the worry of scholars and citizens alike that communicating through a screen instead of face to face largely removes the need to take emotional risks in our relationships.[17] After all, it's easier to plan out what we want to say, share it from a distance, and thereby avoid having to experience the discomfort of facing another person's unfiltered and often unexpected reaction. (Apps, by the way [or, perhaps, not just by the way], are the ultimate filter.)

In our focus groups, we learned that many of today's youth consider it less intrusive to send a text rather than call someone, and it's not uncommon for them to end relationships through text message or Facebook rather than in person. Similar to this is the phenomenon of text cancellations that many of us apparently now rely on to break plans with others at the last minute.[18] Turkle contends that this sort of arm's-length way of conducting relationships ultimately empties them of true intimacy. She warns: "There is the risk that we come to see others as objects to be accessed—and only for the parts we find useful, comforting, or amusing." This emptying out of intimacy is likely what one focus group participant had in mind when she observed tellingly: "Kids are more and more connected, but less and less *really* connected."[19]

There may be another way in which new media technologies remove the vulnerability from our interpersonal relationships and distance us from each other. In a provocative and much debated op-ed, "How to Live without Irony," scholar

Christy Wampole observes a strong ironic sensibility among today's generation of youth.[20] In her rendition, young people wear Justin Bieber T-shirts ironically, watch *Glee* ironically, and give each other birthday gifts ironically. By bathing their actions and interactions in a wash of sarcasm, young people distance themselves both from their actions and from other people. According to Wampole, the Internet supports—indeed, encourages—this ironic turn. Online, the actions of public figures are transformed instantly into derisive memes and circulated widely. The addition of a witty hashtag at the end of a tweet empties it instantly of any seriousness. This sensibility is reinforced nightly on TV—and subsequently posted, shared, and tweeted about online—by Jon Stewart and Stephen Colbert, who wryly ridicule newscasters, politicians, and other well-known personalities. By turning everything into a joke, youth risk nothing because they make nothing of themselves vulnerable. Yet vulnerability is precisely what's needed to connect with other people in an honest and meaningful way.

Distance goes hand in hand with disruption. In every one of our focus groups, participants commented on the disruptive quality of new media technologies. The perpetual buzz of text messages and app notifications (whether friends' Facebook updates, sports scores, or breaking news) issuing from their cell phones pulls youth away from their in-person conversations. In an extreme example, Molly recalled a brunch with one of her dorm mates in which her friend spent the entire time looking down at her cell phone rather than talking to Molly. "I didn't want to just sit there, so I started playing a game on my phone." Incredulous, Howard asked if Molly had at least attempted to make conversation. She had, but none of the topics she offered was able to draw her friend's attention away from her phone. Not wanting to "make the moment awkward," Molly gave up and turned to the safety of her own phone.

Pointing to similar anecdotes from their own observations of youth, our focus group participants expressed concern that mobile technologies and social media threaten to diminish the quality of young people's face-to-face interactions. One participant remarked, "Kids don't have enough practice with face-to-face interaction. They don't go out and play kickball. They don't know how to greet each other."

The disruptive nature of today's technologies came through in our analysis of high school students' creative writing. For each story, we investigated the degree to which technology plays a central role in the plot, as well as any discernible attitudes expressed toward technology by the author or characters. In both the last decade of the twentieth century and the first decade of the twenty-first century, technology and media appear only peripherally in the vast majority of the stories. We did discover an interesting shift in the role of technology in interpersonal relationships, however. In the early stories, technology is never portrayed as disruptive to a relationship; media are even presented in a few of the stories in connection to a shared experience, such as characters reading the newspaper together or watching television news as a family. In several of the more recent stories, by contrast, technology is seen interrupting relationships; indeed, in only one story does the author portray a shared experience related to media (one of the characters is seen watching cartoons with a neighbor). It appears that today's young authors are attuned to the disruptive potential of the new media technologies that pervade their lives.

Among the relationships at risk of disruption by today's media technologies, the family may be particularly vulnerable. On the one hand, as we've noted, families have never been more connected. Thanks to cell phones, instant messaging, and email, discussion is no longer restricted to the morning scramble and the semiregular evening meal but can now extend throughout the entire day. And, as we've seen, when children leave home for college, these technologies allow them to maintain the same level of communication with their parents as when they lived under the same roof.[21] While parents appear generally optimistic about the role of technology in their family life, there seems to be a tipping point. In one survey, parents supported the view that too much technology in the home—too much time online, too many gadgets—has an isolating effect and reduces family time and closeness.[22] This state of affairs resembles the so-called post-familial family, in which families spend more time interacting with their gadgets than with each other.[23]

FROM ISOLATION TO INTIMACY

So much for the isolating effects of an app mentality. As we argue throughout this book, apps can be beneficial if used well. Indeed, evidence accumulated over the past decade suggests that many youth reap considerable interpersonal benefits from their digital media activities.[24] This body of research indicates that, by and large, young people use online communication not merely to substitute for face-to-face communication but rather to augment it. Digital media are thus associated with a *stimulation effect,* whereby the added opportunities to communicate with one's friends translate into increased feelings of closeness to them.

We've identified similar benefits in our own research investigating high school students' experiences with and perceptions of their online peer communications.[25] We find that these online communications can support a sense of belonging and self-disclosure, two important mechanisms through which intimate bonds are formed during adolescence. Digital media may be particularly beneficial for youth who face ostracism in their offline contexts, helping them to find or forge a sense of belonging in a sympathetic community online.[26]

Of course, although a feeling of belonging is preferable to a feeling of isolation, it does not necessarily equate with benign ends (one can belong to a "hate group"—as turns out to have been the case with perpetrators of mass shootings). Nor is a connection necessarily intimate: it may be better described as transactional, rather than as warm, let alone transformational. Consider our discussion of Snapchat in the previous chapter. We noted how the self-destructing messages that people exchange using this app don't support a dialogue so much as a series of one-way dispatches that may lack connection to each other.

Another app, Facetime (Apple's answer to Skype) can also be used to illustrate the ease of falling into a transactional rather than a transformational interpersonal exchange online. When Katie and Molly first talked remotely using Facetime, the first thing Katie noticed was that genuine eye contact is impossible. If you want the other person to feel like you're looking them in the eyes, then you have to look into the camera, not their eyes. In other words, to create the *illusion* of eye contact one must actively *avoid* it. Something else that Katie noticed instantly was her own image in the corner of the screen. She found it hard not to glance over at it periodically, which turned her attention away from Molly and onto herself. Apparently, Molly was equally, if not more so, enticed by the "Narcissus trap." In fact, at one point in their conversation, Katie was confused when she made a funny face but Molly didn't react in the slightest. When Katie called her on it, Molly admitted somewhat sheepishly that she'd been focusing on her own image and facial expressions instead of her sister's. Overall, Katie's experiences with Facetime, Skype, and Hangouts on Google have led her to conclude that, while it's great to be able to connect with others across distances, it's difficult—if not impossible—to achieve the level of deep, warm connection that face-to-face contact provides.

Ultimately, whether digital media lead youth to feel connected to or isolated from others will depend on their orientation toward these media: Is theirs an app-enabling or an app-dependent stance? Do they use apps to augment or replace their offline relationships?

EMPATHY LOVES (AND NEEDS) COMPANY

Isolation is an individual-level problem, but one that can have larger social effects, straining empathy and diluting prosocial attitudes. Considerable evidence suggests that today's young people are less empathetic than youth of the 1980s and 1990s. Researchers at the University of Michigan came to this conclusion after analyzing the combined results of seventy-two studies of American college students conducted between 1979 and 2009.[27] They found a small but significant decline over time in the number of students who agreed with such statements as "I sometimes try to understand my friends better by imagining how things look from their perspective" and "I often have tender, concerned feelings for people less fortunate than me."

Other trends parallel this decline in empathy and perhaps serve as indicators of it. The Michigan researchers themselves point to research showing an increase in crimes against stigmatized and marginalized groups, such as the homeless, Hispanics and perceived immigrants, and lesbians, gays, and bisexual and transgender individuals. There's also evidence that sexual harassment and stalking on college campuses have increased in recent years.[28] If we consider these disturbing trends in light of the empathy decline, we might posit that people are more likely to harm others when they lack the ability to see themselves in other people. Indeed, the absence of empathy is a trademark of the sociopath.

The decline in empathy and rise in hate crimes seems at odds with our discussion in the last chapter of youth's greater acceptance of people who are different from them. With respect to hate crimes, it's important to note that most young people do not commit such acts. The increase in these crimes involves a relatively small proportion of individuals who may be disproportionately affected by the general decline in empathy. With respect to the apparent incongruence between increasing acceptance of difference and decreasing empathy, it's worth noting that acceptance and tolerance of others is not the same thing as putting oneself in their shoes. Furthermore, recall that our focus group participants observed a certain shallowness in youth's approach to people, practices, and cultures that differ from their own. In other words, their acceptance doesn't seem to be accompanied by a greater understanding of others. Nor is it the case that, given a choice, young people voluntarily spend time with those from different racial or ethnic groups. Silo-ing appears to be alive and well in many high schools and colleges.

In our own analysis of seventh- and eighth-grade fiction, we identified a decrease over time in the number of stories in which the author featured characters who differed notably from him- or herself. In the mid-1990s, 32 percent of the stories depicted a main character who differed from the author in terms of gender or age. Not one story from the late 2000s did so. This decline in "character play" indicates that today's students may be less inclined—less able?—to assume the perspective of characters who are different from them. And if one can only compare oneself with individuals already in one's social group or (more likely) to their idealized portrait on a social network, there's little wonder that greater empathic stretches are undermined.

The Coarsening Effect of Digital (and Predigital) Media

In the Michigan study, the greatest drop in college students' empathy scores occurred after 2000. It's hard not to consider this trend in light of the explosion in social media that took place during the same period. Could viewing the world through our apps be hurting our ability to view the world through another's eyes?

To explore this question, we turn first to a 2011 Associated Press–MTV poll that suggests that online speech may have a coarsening effect on the way people relate to each other.[29] In the poll, 71 percent of fourteen-to-twenty-four-year-olds said people are more likely to use racist and sexist language online or through texting than in person. Molly wasn't surprised by this statistic. In her experience, people are generally meaner online than in person. "I think kids my age find it easier to make fun of someone through a veiled post on Facebook or Twitter. I think they forget who they are online and use [their online profile] as a separate identity almost that loses responsibilities and is invincible to consequences because it's just

black ink on a screen." She explained that the public pages of
Facebook can be sites of particular cruelty. "On comments or
wall posts people can unleash their meanness, in which case
the thread can turn into a public 'debate' where friend groups
take sides and people join in." Molly noted that online cruelty
is especially common among middle school students and girls,
an observation that's consistent with existing evidence.[30]

Molly also explained how photos can be used to embarrass
another person publicly, particularly someone who's considered weird or uncool. As an example, she recounted a story
from her own middle school experience. Soon after she joined
Facebook, she came across a photo album posted by one of
her classmates. "I was in a few of the pictures, which had been
taken in 2006 or 2007 (so before I met the glories that are
tweezers, contact lenses, and braces) and one of my supposed
friends (at that time) had commented, 'Thank God for braces
and contacts,' which was met by a series of agreements and
similar comments between the other people in the album. I
completely agree with them, but to see it written on Facebook
in such a mocking way was harsh." Though Katie can remember experiences of bullying from her youth, a notable difference between her experiences and Molly's is that she could at
least put them on pause when she came home after school.
Now, in a manner reminiscent of the former British Empire,
"the sun never sets on bullying."[31]

Though this sort of online bullying is somewhat more prevalent among girls, boys aren't off the hook when it comes
to the coarsening effects of digital media. Sexual harassment

is common in certain online gaming communities, in which women are called various derogatory names, offered virtual money in exchange for online sex, and, stalked, both online and off. In one heinous example, a male gamer responded to the efforts of one woman to combat sexual harassment in online games by creating his own game in which players punch the woman's virtual image, adding bruise upon bruise, until the screen turns red.[32]

Even more ubiquitous than online games is online pornography. Pointing to the unparalleled access that today's youth have to pornographic material, some of our focus group participants expressed concern about the emotional effects of experiencing pornography as a dominant model of relationship. In particular, they worry that boys will approach their romantic relationships by sharing less of themselves and making less effort to understand and connect with the emotional life of their partner. Adolescent males will come to expect that their sexual partners will be as willing and as undiscriminating as is the composite porn star.

Scholars have suggested a link between youth's consumption of online pornography and a "hook-up" culture that's arisen over the past fifteen years among American high school and college students.[33] One educator we interviewed said that the teens she works with today seem to consider oral sex "less personal than kissing." In one study, researchers found that today's college students are hesitant to enter into committed relationships, preferring instead to cycle through a series of casual relationships based on sex instead of romance.[34] (This cycling is made all the more easy with dating and hook-up apps.) The study's authors suggest that it's not that today's youth aren't interested in romance: they are. However, their fear of making themselves vulnerable to another person outweighs their desire for romance. For these youth, a series of isolated hook-ups feels less risky than a sustained emotional attachment to another person. In a similar vein, and harkening back to our earlier discussion of the "whatever generation," one therapist observed: "The goal is to feel nothing. . . . In an overstimulated world, there's a 'cool thing' about dissociating and feeling nothing. As if that would be the goal of a sexual encounter—to be able to walk away from it saying, 'No big deal. I'm empowered.'" Howard was amused—or perhaps bemused—to learn that many young people hook up first, then consult the relevant Facebook entry to decide whether they want to see their sex partner again "in the light of day."

Our focus group participants also implicated today's suite of reality television shows in the degradation of young people's interpersonal relationships. Shows like *Jersey Shore, Bad Girls Club,* and the *Real Housewives* series feature "real-life" characters who consistently act contemptibly toward one another. In each season of *Bad Girls Club,* for instance, seven self-described "bad girls" are put together in a mansion and then filmed as they proceed—predictably—to behave badly. The camera follows them around the house, pool, and limo as they push, slap, and scream at each other and generally do everything they can to elevate their own status in the house at the expense of their housemates. This type of behavior evokes images from tabloid talk shows dating back to the early 1990s, like *The Jerry Springer Show* (which, by the way, was still airing new episodes as of the writing of this book). However, the increase in the number of such shows and the ease with which they can be accessed—via the Hulu, You-Tube, or Netflix apps—seems to have expanded their imprint on the cultural psyche.

The crazy-quilt nature of online communications among youth was brought home sharply to Howard when he eavesdropped (with permission) on a scheduled conversation about digital life among a dozen teenagers. For the first few minutes of the discussion, the young people talked about social network communications in a mild and benign way. But then one young man mentioned how reputations can be destroyed. This simple remark opened up a floodgate of testimony, with several students detailing ways in which classmates and friends had been lambasted and bullied, often with no defense from their supposed friends. It was as if a curtain had been lifted, revealing the backstage events on Facebook. As another instance of the schizoid quality of online communications, several young people in our studies reported the strange situation in which a person reveals the most intimate details online, subsequently reverting to perfunctory, distanced interactions offline with the very peers to whom the messages had previously been directed.

Though illuminating, Molly's experiences and the observations of our focus group participants are more suggestive than proof positive that today's media landscape adversely affects social relations. As with most social science research, it's not easy to establish a cause-and-effect relationship between two complex variables. However, we did come across one study that moved in this direction. The researchers designed a series of clever experiments to test whether cell phone use affects college students' prosocial tendencies.[35] In one experiment, participants completed a questionnaire that measured their willingness and motivation to engage in a variety of prosocial acts. Half of the participants were asked to use their cell phones for a brief time before filling out the questionnaire. These participants were less likely than participants in the control group to say that they would volunteer for a community service activity. They were also less likely to persist in solving word problems that they were told would result in a monetary donation to a charity. Remarkably, these differences remained even when the participants in the treatment group were asked to draw a picture of their cell phone and then think about how they use it.

To explain their findings, the study's authors suggest that the cell phone enhances participants' feelings of social connectedness, thereby decreasing their need to seek social connection elsewhere. The implications of this influence are profound. Think about who you communicate with most via cell phone—more likely than not, close family and friends dominate the list. It may be that our cell phone use diminishes our inclination to seek social connection beyond our close circle of intimates.

A similar dynamic may be playing out on the Internet. In *The Filter Bubble*, Eli Pariser explains how search engines and social network sites show us only what we want to see (or what they think we want to see).[36] He uses Facebook's Edge-Rank as one example of how this works. EdgeRank uses an algorithm to rank users' friends list according to how much interaction each user has with each person on the list. EdgeRank then uses that ranking to structure users' newsfeeds so they see more from the friends at the top of the list. Google's search algorithm works in a similar way such that two people conducting an identical Google search (whether it's "performing arts in Atlanta" or "2012 presidential election") will be shown a different set of results based on what Google knows about them (and drawing on previous search history, Gmail contacts and exchanges, YouTube posts and viewing habits—Google knows a lot!). Pariser argues that such algorithms have a siloing effect, causing us to encounter only like-minded people and ideas online. It's difficult to empathize with perspectives that we never see.[37]

TAKING STOCK

Apps are, ultimately, shortcuts. We've seen in this chapter a number of shortcuts in how today's young people carry out their interpersonal relationships. These shortcuts make interacting with others much quicker, easier, and less risky. If used in moderation and to augment rather than replace face-to-face contact, such conveniences can certainly enable meaningful relations and, at their best, strengthen and deepen personal bonds.

However, convenience comes at a cost. We've considered the extent to which particular features of mediated communication may underpin the increasing isolation and declining empathy identified by a range of scholars. In light of our discussion in the previous chapter, we're particularly attuned to the role of diminished risk-taking in these societal trends. It may feel more comfortable to remove the risk from social interactions, but if we don't put ourselves out there, we can't truly connect with others (isolation). And, if we don't truly connect with others, we can't put ourselves in their shoes (empathy).

Combining the two areas we've surveyed so far produces a picture of the prevalent consciousness of members of the App Generation. When it comes to a sense of identity, many members of this generation feel pressure to proceed down a path that is valorized by society (other-dependent) and that promises them the life and career they feel they deserve (app-dependent). When it comes to a sense of intimacy, such young people make ready and skilled use of modes of connection that are instantly available but with the concomitant failure to pursue riskier, yet potentially more meaningful relationships with one another. Only those young people able to resist the Narcissus trap and the Circean lure of the apps-of-the-moment are likely to form a meaningful identity or to forge intimate relationships with others. Having surveyed young people's sense of self and their relation to others, we turn next to the kinds of imaginative worlds that they create and how these may be shaped by the digital tools at their disposal.

SIX

Acts (and Apps) of Imagination among Today's Youth

APPS LIKE SKETCHBOOK, BRUSHES, ArtStudio, Procreate, and ArtRage allow artists to draw, sketch, and paint using their smartphone or tablet. Photographers can create and manipulate images with Flixel, Instagram, Fotor, and PhotoSlice. For aspiring filmmakers, there's Viddy, iMovie, Video Star, and Movie360. Musicians can compose and arrange their music using SoundBrush, GarageBand, Songwriter's Pad, and Master Piano.

We could make similar lists for just about any artistic genre. A sizable portion of the app ecology is devoted to supporting artistic production. Even apps that aren't ostensibly meant for creative pursuits lend themselves to imaginative uses. Recall our earlier discussion of the mini-performances that Molly stages and sends to Katie via the messaging app Snapchat. Living up to the name that we have bestowed on them, apps have changed how members of the App Generation engage their imaginations. We'll explore what's gained and what's lost by using apps (and other digital media) for the purpose of artistic expression.

We find that digital media open up new avenues for youth to express themselves creatively. Remix, collage, video production, and music composition—to name just a few popular artistic genres of the day—are easier and cheaper for today's youth to pursue than were their predigital counterparts. It's also easier to find an audience for one's creative productions. The app metaphor serves us well here, since apps are easy to use, support diverse artistic genres, and encourage sharing among their users.

And yet—reflecting patterns we observed in youth's expressions of personal identity and experiences of intimacy—an app mentality can lead to an unwillingness to stretch beyond the functionality of the software and the packaged sources of inspiration that come with a Google search. We ask: Under what circumstances do apps enable imaginative expression? Under what circumstances do they foster a dependent or narrow-minded approach to creation?

Before proceeding, two background points. First, a word about our focus on art. We recognize that the imagination can be exercised in just about any sphere, be it science, business, a hobby, or sports. Indeed, Erik Erikson was referring to a wide range of endeavors when he wrote about the challenge of using one's mind and one's resources actively and imaginatively to pursue a meaningful, generative life. We focus on art because it may have the broadest provenance, because it's generally what people think of first when they think of imagination, and because we have had the opportunity to examine wonderfully revealing collections of art secured over a two-decade span.

Second, a word about the term *imagination*. We are interested in how young persons use their cognitive, social, and emotional capacities to broaden their understandings and enrich their productions—to think outside the box, as the saying goes. Many commentators are interested in this twenty-first-century skill, and they readily invoke words like "creativity," "innovation," "originality," and "entrepreneurship" to capture the idea. We like "imagination" because it focuses sharply on the psychological process that the young person can bring to an activity—and, to be candid, because it allows us to speak of the Three Is.

FROM VIDEOTAPE TO VIDDING

There can be little doubt that apps and other digital media technologies have altered the landscape of imaginative expression. They've affected virtually every facet of the creative process, encompassing who can be a creator, what can be created, and how creations come into being and find an audience.

Let's consider a few examples. If you've watched the Super Bowl (or, more important, the commercials) at some point over the past several years, you've likely seen one or more of the consumer-generated ads for Doritos. PepsiCo Frito-Lay, the company that sells Doritos, initiated the "Crash the Super

Bowl" advertising campaign in 2007 by inviting fans to design their own thirty-second ads for Doritos and to vote online for their favorite finalists. After submitting their projects, many of the amateur filmmakers used social media platforms like YouTube and Facebook to build support for their commercials. For the 2012 Super Bowl, submissions topped 6,100 (the highest ever) and online voting reached into the hundreds of thousands. One of the competition winners—featuring a baby and grandmother duo who team up to snatch a bag of Doritos from a playground bully—scored the no. 1 spot on the USA Today/Facebook Super Bowl Ad Meter. The Ad Meter itself marked the first time that USA Today partnered with Facebook to allow viewers—rather than preselected, "authorized" panelists—to choose their favorite Super Bowl commercial. The Ad Meter win earned the ad's creator, a former special education teacher, a bonus prize of one million dollars from PepsiCo.

To be sure, the "Crash the Super Bowl" campaign is partly a story about how corporations are finding ways to use social media to grow their profits. But it also illustrates how today's digital media technologies are shaping the creative process in new ways. The introduction of inexpensive and flexible video-taking hardware (smartphones, digital cameras, tablets) and video-editing software (many of them available as apps like iMovie, Viddy, and Movie360) has both lowered the bar for entry into filmmaking and elevated the quality of amateur productions. The advent of social media is also transformative.[1] Scholars talk about the important role that "the field"—essentially, those who judge any given work—plays in the creative process.[2] Video-sharing sites and apps like YouTube, Vimeo, and Facebook have dramatically expanded the size of this field, as well as the amateur videographer's access to it.[3]

Molly is one such amateur videographer who's taking advantage of the creative opportunities introduced by digital media technologies. She started making videos at age eleven with the iMovie software that came standard with her first laptop, a MacBook. She found the software intuitive and enjoyed tinkering with the various effects, title sequences, and music overlays. Her movies have typically involved clips of time spent with friends and family, artfully pieced together and set to evocative music. She's even posted a few on YouTube. "While they get nowhere near the millions of views that a Justin Bieber video gets, it's nice to think my teeny videos have been watched by other people."

We've interviewed several other young creators during our research and found them to be using digital tools in a variety of imaginative ways. Nineteen-year-old Danielle told Katie about her experiences creating and sharing vids on the online journaling community LiveJournal. The young videographer described vids as short videos consisting of clips from a television show or movie and set to popular music. As such, they belong to the same family as musical mashups, made popular by the TV show *Glee*. Typically, there's a strong fan community surrounding the chosen television show or movie. Viewers are expected to draw on their background knowledge of the original work to interpret the message conveyed by the vidder's scene and music selections.

The majority of Danielle's vids feature scenes from the sci-fi television series *Stargate Atlantis,* though more recent vids draw from movies. In addition to expressing her imagination, Danielle uses her vids to convey political messages, typically of a feminist bent.

The vidding community that Danielle belongs to on Live-Journal represents an important part of her vidding experience. There, she learns technical skills from more seasoned vidders, receives constructive feedback on the work she shares, and offers her own feedback on others' work. This community isn't limited to the virtual world. Days after her first interview, Danielle flew to Chicago with her best friend to attend a vidding conference where she was excited to meet some of the more "famous" vidders who have mentored her on Live-Journal.

A variety of other social media platforms besides Live-Journal—many of which are available in app form—give youth the opportunity to share their creative work with others. Figment is a "social network for young-adult fiction"—in the words of cofounder Jacob Lewis, a former managing editor of the *New Yorker*—where teens can share their creative writing with other teen authors. Like other social network sites, teens create a profile page that includes a profile picture, a self-description, a list of followers, group memberships (such as "The Poets of Figment" and "Queer Figs"), and a wall for other users to write comments. Given the literary focus of the site, profile pages also include a list of favorite authors, links to the profile owner's original writings as well as reviews about other writings, and the array of badges that the teen has earned from his or her participation on the site. (A few examples: the "Wordsmith" badge, awarded when a teen has posted ten original pieces; the "Bookworm" badge, awarded when a teen has read twenty-five writings by other teens; and the "Critic" badge, awarded when a teen has written thirty reviews). DeviantART, another site with a similar setup, focuses on visual art instead of creative writing. Sites like these open up exciting possibilities for youth creators to share and receive feedback on their work.

There's much to be excited about in these examples of creative expression in the digital era. Scrolling through the writing and art on sites like Figment and deviantART, it's clear that many young people are using these digital tools to exercise their imaginations. And yet, one wonders what sort of exercise they're getting with these tools. Take Danielle's vids as a case in point. There are certainly those who celebrate vidding and other forms of remixing as original and creative acts of artistic expression.[4] Others, however, argue that there's nothing original about reusing work created by others.[5] Of course, this argument existed before the arrival of digital media technologies—think of Marcel Duchamp's toilet "fountain" or Andy Warhol's soup cans. The argument is simply brought into sharper relief in this copy-and-paste culture.

There's also a question of whether the constraints built into apps and other computer software adversely restrict the creative process. Thinking back to our opening example of young children being introduced to a new toy, we raise the question of whether children are better off making up their own games in the backyard or playing video games designed by professional game designers.[6] Even when media aren't part of the creative act itself, one wonders how young people's steady diet of text messages, Facebook updates, tweets, and streaming music affects their ability to engage deeply in the creative process and, sooner or later, strike out on their own.

IMAGINING, THEN AND NOW

Before we consider how apps and other digital media affect imagination, let's first explore whether young people's imaginative processes have actually changed since the introduction of new media technologies. Imagination is a difficult concept to define, let alone measure. Nonetheless, psychometricians have given it their best effort, typically by administering various tests of creativity. Perhaps the most widely used creativity test is the Torrance Test of Creative Thinking (TTCT). Developed in 1966 and currently used worldwide, the TTCT measures several dimensions of creative potential, including intellectual curiosity, open-mindedness, verbal expressiveness, and originality. Though not without its critics, the TTCT has been found to predict creative achievement better than other standard measures of creative or divergent thinking.[7] Empirical evidence suggests that high scores on the test successfully predict subsequent creative careers and accomplishments.[8]

In a widely publicized study, Torrance scores from approximately three hundred thousand children and adults were used to investigate whether the creativity of Americans has changed over the preceding twenty years.[9] The research documents a pronounced decline in scores across all areas of the figural test. The largest drop was seen in scores on *elaboration,* which includes the ability to elaborate on ideas and engage in detailed and reflective thinking, as well as the motivation to be creative. Declines were also found in *fluency* (the ability to generate many ideas), *originality* (the ability to produce infrequent, unique, and unusual ideas), *creative strengths* (which include emotional and verbal expressiveness, humorousness, unconventionality, and liveliness and passion), and *resistance to premature closure* (the inclination to remain open-minded, intellectually curious, and open to new experiences). Overall, the declines were steepest in more recent years, from 1998 to 2008, and the scores of young children—from kindergarten through sixth grade—decreased more than those of other age groups.

Curiously, in the same issue of the journal in which this study was published, another group of researchers published an article whose findings paint a more optimistic picture of changes in youth creativity.[10] The researchers investigated changes in the pretend play ability of children between the ages of six and ten during a twenty-three-year period. Though not synonymous with creativity, pretend play has been found to predict divergent thinking, which is itself a marker of creativity.[11] The researchers analyzed results from fourteen studies conducted between 1985 and 2008. Each study used the same instrument—the Affect in Play Scale (APS)—to measure the pretend play of children in grades one through three. The APS measures multiple dimensions of pretend play, including *imagination* (How many fantasy elements and novel ideas does the child produce?), *comfort* (How comfortable is the child engaging in play, and how much enjoyment does he or she experience?), *organization* (What is the quality and complexity of the play plot?), *frequency* and *variety of affect* (How often does the child express emotion, and what is the range of emotions expressed?), and *positive* and *negative affect* (How often does the child express positive and negative emotions?).

Of the seven play dimensions measured, only *imagination, comfort,* and *negative affect* showed any significant change over the twenty-three-year period. *Imagination* and *comfort* both increased significantly, suggesting that young children have become more imaginative in their pretend play and have come to derive greater enjoyment from play. In contrast, *negative affect* during play decreased over time. This last change is the only finding that accords with the other study showing declines in creativity, since negative emotional themes in children's play have been linked to divergent thinking.[12] At the end of their article, the authors of the play study acknowledge the inconsistency between the main findings of the two studies, though they offer little in the way of resolution beyond the familiar call for "more research."

WHAT OUR DATA SAY ABOUT
CHANGES IN CREATIVITY

The third set of research findings we'll consider involves our own investigation into changes in youth creativity that may have occurred over the past twenty years. Rather than look at scores on tests of creativity or its correlates (like play), we chose to examine the actual creative productions of young people. This approach provides a more naturalistic view into young people's creative processes. To that end, we conducted an extensive analysis of short stories and visual art created by middle and high school students between 1990 and 2011. (In our methodological appendix we detail how we analyzed these works, including the steps we took to ensure that our classification of each piece was done in a consistent, objective manner.)

It would be appealing if our investigation could decisively resolve the apparent conflict between the two studies discussed above. Unfortunately, such is the nature of research in this minefield-packed area that we're unable to do so. In fact, our findings actually complicate the story further, but in ways that we consider instructive and revealing.

Part of our investigation involved an extensive analysis of 354 pieces of visual art published over a twenty-year period in *Teen Ink,* a national teen literary and art magazine based in Newton, Massachusetts. Our analysis revealed a notable rise in the complexity of artwork published between 1990 and 2011. For example, we analyzed the background of each piece, evaluating how the artist treats the space around and behind the foreground figures and objects. Compared to the early pieces, the backgrounds in the later pieces are more fully rendered. In other words, figures are more likely to be situated in a fully completed context in the later pieces, whereas the objects of the foreground are more likely to be seen floating in blank or partially rendered spaces in the early pieces.

This difference was large: 78 percent of the later pieces were categorized as fully rendered, compared to only 49 percent of the early pieces. As a consequence, one experiences the later pieces as more fully developed and complete than the early pieces.

Another marker of complexity that we examined pertained to the composition, or balance, of each piece. In particular, we looked at the positioning of the figures and objects in the visual plane: Are they in the center or off to one side? The number of centrally composed pieces dropped from 58 percent of the early pieces to 49 percent of the later pieces, suggesting that the contemporary artists are somewhat more likely to experiment with the location of their figures on the visual plane.

We also looked for evidence of cropping: Do the figures extend beyond the visual plane? Here we found a rise in the number of cropped pieces, from just 4 percent of the early pieces to 15 percent of the later pieces. Again, the contemporary artists appear to be more comfortable presenting their figures in a less conventional way than the earlier artists.

This departure from convention is also represented by our analysis of the production practices employed by the artist. Not surprisingly, the number of pieces that were manipulated through digital means (Photoshop, postproduction photography manipulation, and so on) increased markedly over the twenty-year period. Less than 1 percent of the early pieces display evidence of digital manipulation; that percentage increased to 10 percent for the later pieces. The later pieces also depart from traditional production practices with respect to the range of media represented. For instance, the number of pieces that employed either traditional pen and ink or another form of drawing (for example, charcoal or pencil) declined from 55 percent of the early pieces to 18 percent of the later pieces. Conversely, the number of pieces using less traditional media—such as digital art, collage, public art, found objects, and mixed media—rose from less than 1 percent of the early pieces to 9 percent of the later pieces.

Our final evidence for the increasing complexity of teen artwork concerns the overall stylistic approach that the artist employed. Examining the pieces holistically in terms of both content and technique, we classified each one as falling into one of three categories: *conservative, neutral,* or *unconventional.* We classified a piece as *conservative* if it followed the traditional conventions of its medium in an appropriate way and didn't deviate from conventional practices in either technique or content. We classified a piece *neutral* if it neither followed traditional art genres/styles nor offered a unique or provocative take on its subject. If the piece demonstrated obvious provocation in either content or technique, we classified it as *unconventional*. Technically unconventional pieces might play with perspective or use media in unusual ways. Work considered to have unconventional content might depict figures in unlikely contexts—bodies climbing out of trash cans, abstract forms made up of screaming, disembodied heads. Sometimes a piece was considered unconventional in both content and technique, sometimes just one or the other. Our analysis revealed that the percentage of *conservative* pieces declined from 33 percent of the early pieces to 19 percent of the later pieces, whereas the number of *unconventional* pieces rose from 19 percent to 28 percent.

This departure from the expected suggests a growing sophistication in the art produced by young artists over the twenty-year period of our investigation.

Our analysis of teens' creative writing—both among middle school students and high school students—produced a notably different pattern of changes. For instance, when evaluating the types of genre employed by the high school writers (for example, science fiction, fairy tale, or historical fiction), we found evidence of a decline in what we call "genre play." A story was deemed to display genre play when it deviated from a traditional realist perspective, typically by incorporating fantasy elements such as magic or absurdist themes. Among the stories written in the early 1990s, 64 percent incorporate such fantasy elements. By contrast, nearly three-quarters of the later stories (72 percent) evidence no sign of genre play at all (for sample stories from each time period, see pages 136–139).

We found a similar, though slightly less pronounced, trend in our analysis of middle school fiction. Although we categorized the majority of the pieces in both sets as realism, about one third (32 percent) of the early pieces exhibit genre play whereas, in the later set, only one tenth (10 percent) of the pieces contain any elements of genres other than realism.

A more notable distinction emerged when we looked at the plot of each middle school story. For this analysis, we noted whether the plot was meandering or fast-paced; examined the central conflict of the story; and documented significant moments of rising action. From this analysis, we identified three dominant categories of plot: *every day, every day with a twist,* and *not every day. Every day* stories have fairly mundane plots; they tend to describe events at home and at school that could take place on any day, at any time of the year. We defined *every day with a twist* as a plot that is mostly familiar or mundane but that contains at least one moment of rising action that could not happen every day. *Not every day* plots contain significant fantastical elements and/or impossible occurrences. We found that the same number of stories in both early and later sets was classified as *every day with a twist* (27 percent in each set). However, there was a marked shift between early and later stories away from *not every day* stories toward *every day* stories. Nearly two-thirds of the early stories (64 percent) were classified as *not every day,* whereas only 14 percent of the later stories were so classified.

Similar patterns emerged when we examined other story elements like setting, time period, and narrative linearity. For instance, in the high school data set, the early stories are more likely to follow a nonlinear story arc; the later stories tend to unfold in a conventionally linear manner. Whereas only 40 percent of the early stories were classified as linear, fully 64 percent of the later stories were so categorized.

Among the middle school stories, we found that the early stories were more likely to be set in an unfamiliar (at least to a middle school student) location, such as a World War II battle. Whereas almost a third (32 percent) of the early stories took place in distant locales, only one of the later stories (5 percent) involved an unfamiliar setting. Paralleling this trend, we also found that the time period of the early stories was more likely than the later stories to be different from the time period in which the story was composed.

Considered together, these changes in genre, plot, story arc, setting, and time period suggest that, while teens' visual art has become less conventional over time, creative writing emanating from this age group has become more so.

One last noteworthy change that we identified in the high school stories concerns the formality of the language employed by authors. Compared to the early stories, the language in the later stories is considerably less formal. The contemporary authors are more likely than their counterparts from the early 1990s to incorporate expletives ("piss," "shit"), slang ("awesomeness"), and made-up words ("glompy," "smushed") into their prose. The difference is dramatic. Only 24 percent of the early stories include informal, pedestrian language, whereas fully 80 percent of the later stories do so. In short, the early stories may be more "out there" in terms of their incorporation of magical and absurdist themes, but the language they use to depict these fantasy worlds is somewhat less flavorful.

This high school story, from the early 1990s, includes fantasy elements, artful word choice, a range of references, and figures of speech.

The Psychiatrist

Now I must see my psychiatrist, Dr. Sanborne. How I hate these weekly visits! The notion that anything is wrong with me is absurd, of course. These visits merely erode my checkbook.

As I step into his office, Sanborne scuttles sideways out from under his mahogany desk to greet me, as usual. His blue shell, encrusted with tiny jewels, sparkles, and his fragile feelers begin tracing an invisible diagnosis in the air.

"Good morning! How are you feeling? I think we are making great progress in our sessions. Please sit down," he says, in a voice like sand being sifted.

"I am perfectly fine." I lie down on the leather patients' couch reluctantly. From my vantage point I can see only his eyes, two hypnotist's orbs, waving on their stalks. I decide this is the last session I will attend. After a pause I blurt out, "Look, Doctor, this is useless. You

know my mind is as round and perfect as a seashell."
His pincers are clicking rapidly, like a machine analyzing my responses.

"You forget," Sanborne replies, "that there is always
an opening through which I can crawl into the hopeless
spiral of your subconscious." He climbs onto the couch.
"Listen, do you not hear the rustling ocean of insanity,
splashing on the walls of this very room?" He is very sly,
but I will not let him trick me this time.

"Nonsense! You are the ridiculous one. You're not
even a man, just a crafty old crab, greedily snatching
my money and scurrying off to line your burrow with
it," I shout, pulling a pair of redhandled tongs from my
coat pocket with an exaggerated flourish. "I've got you
now, though." I firmly grasp Sanborne by his middle,
and avoiding his furious pincers, thrust him into my
briefcase. "Tonight I dine on boiled crab!"

As I walk out, Sanborne shrieks out from the dark
depths at my side: "Release me! You don't understand
the torment of the psychiatrist's existence: like a doomed
Proteus, I am helplessly transformed by every madman's
delusion!"

Dolsy Smith

This high school story, from the late 2000s, features ordinary language, mundane subject matter, and an easily recog-
nizable descriptive genre.

Age
Ryanne Autin

At home every day, it becomes difficult to not just lounge on the couch and smoke cigars while your wife is not home. It becomes difficult to not watch football game after football game and not change the same white shirt that hides your stomach which protrudes over a pair of pajama pants that you have owned for years. However, the-used-to-be loose elastic band now has tightened. You pull at it periodically to measure the amount of fat you could possibly gain before needing new pants. You are disappointed every time because it always ends up being less than an inch. Luckily, your wife has also left a post-it note on the bathroom mirror with the scribble, "Don't forget to walk the dog, do the dishes, and take the trash out. Making fish for dinner, smiley face, love you!" You are considered an old man now, dressed in a business suit each day, even though you are only going to the store, to the barber, to tend to matters regarding your dying mother. This is the first year of your adult-hood that you have not spent working. Your wife wakes up next to you at six in the morning, seven days a week. Your daughter and son no longer call daily but rather, you receive a text message every other day on a phone with a touch screen that you're not sure how to end a call on. The messages relay pictures of the grandchil-dren in Ralph Lauren, Armani Baby, and Jotum. You worked hard to raise the children well, they married wealthy companions that they may or may not love, and the grandchildren are spoiled, just not like the grilled cheese sandwich you found in the refrigerator last week. No longer within miles of you, your family has become extended, but you hold your hand to your heart before you go to bed every night and say, "Thank you. Thank you. Thank you."

WHAT TEACHERS SAY

To complicate matters further, we also spoke with art teachers (visual art, music, and performing arts) who'd been teaching for at least twenty years and therefore could reflect on changes they've observed in students' imaginative processes over time. Though these teachers celebrated the broad range of creative opportunities now open to today's youth (which we discuss in greater detail below), several arts educators observed that today's students have more difficulty in coming up with their own ideas; they're far more comfortable engaging with existing ones. One participant reflected: "Some of the most artistically skilled kids cannot come up with an idea. They've got full scholarships to Mass Art [Massachusetts College of Art and Design] and they can't come up with an idea. . . . They go to their laptop first. . . . I find that I'm constantly shoulder to shoulder asking what do you see? What does it mean? . . . They're thinking too much or saying 'I have nothing.'" Moreover, when they do come up with their own ideas, they often have difficulty executing them, particularly in the absence of clear "executive assistants." Said another participant, "Before, they used to jump in and see where the materials would take them, now they ask what to do."

The camp directors made a similar point when they reflected on the changes they've seen in skit night, a camping tradition in which campers form groups and perform a short skit for the rest of the camp. Today's campers are more likely to re-create an episode from a favorite TV show than they are to invent their own story, as campers did in years past. The re-creations may be more polished, but the invented stories were more interesting (and more promising) for being original. These observations from the art teachers and camp directors align with the overarching concern expressed by our many informants drawn from different sectors: youth of today display less willingness to take risks in their creative productions.

One of the theater directors told us that both students and student productions are more conservative today. Twenty-seven years ago, his students produced a "very edgy" version of *Alice in Wonderland* that featured a surrealistic set and unorthodox lighting. This year, his students produced the same show using the same script. The more recent version of the show was "cute and sweet," the director lamented, and the students, while talented, did not discern the piece's more subtle political messaging. Today's students are also concerned about potentially "getting into trouble" for mounting productions that might be construed as "provocative." It was not clear with whom the students might "get into trouble"—parents, administrators, peers, or another group of concerned individuals. Lamented one participant, "[They're] rule followers to a fault." These findings echo the search for "correct answers" and "documented procedures" and precise scoring rubrics that Howard has noted among his students in recent years.

REMIXING THE IMAGINATION FOR A DIGITAL ERA

Our participants pointed to the Internet's vast supply of packaged sources of inspiration to explain why today's young people have greater difficulty coming up with their own creative ideas and instead prefer to engage with existing ones. After all, it's easier to reach for Google instead of scouring one's imagination for a new idea. One participant reflected: "The classic example is 'Go outside on a snow day.' How many kids are actually outside making a snowman or having a snowball fight? Not many. They're inside creating a snowman on the computer." Still, one could argue that if the end result is judged to be a creative piece of work, who cares where the source of inspiration came from? Surely one can engage with existing ideas in a creative way.

The trouble is, there's considerable debate over the value of what youth create with these existing ideas. While many respected authorities celebrate remix culture, others are less sanguine.[13] In his book *You Are Not a Gadget,* computer scientist and cultural critic Jaron Lanier bemoans the effects of remix on individual creativity: "Pop culture has entered into a nostalgic malaise. Online culture is dominated by trivial mashups of the culture that existed before the onset of mashups, and by fandom responding to the dwindling outposts of centralized mass media. It is a culture of reaction without action."[14]

Even if a person made a concerted effort to act instead of react, to invent instead of remix, Lanier argues that digital media would still throw up obstacles to creativity. He uses the expression "lock-in" to describe the limited range of actions and experiences open to users when they interact with computer software. As a result of a programmer's (often arbitrary) design decisions, certain actions are possible—indeed, encouraged—while others don't even present themselves as options.

Lanier's primary example of lock-in involves MIDI, a music software program developed in the 1980s to allow musicians to represent musical notes digitally. Because its designer took the keyboard as his model, MIDI's representation of musical notes doesn't encompass the textures found in other instruments, such as the cello, flute, or human voice. Lanier argues that something important is lost when one makes explicit and finite an entity that is inherently unfathomable (or, to invoke another lexical contrast, when one seeks to render as *digital* what is properly seen as *analog*). Moreover, since MIDI was an early and popular entrant into the music software industry, subsequent software had to follow its representation of musical notes in order to be compatible with it. As a result, the lock-in was reified. MIDI is a good example of how early design decisions can circumscribe subsequent creative acts.

Apps may represent the ultimate lock-in. Consider the Songwriter's Pad app, a tool for writing songs and poems on an iPad. This app is intended to make songwriting easier by breaking down the process into manageable sections and helping the songwriter to keep track of his or her ideas and progress. It's also supposed to inspire creativity by supplying built-in sources of inspiration such as a rhyming dictionary and thesaurus. In addition, the app generates words or phrases based on specified moods like anger, desire, love, or hope. For instance, clicking the "anger" button calls up such phrases as "you ripped my heart out" and "stormed off fuming." Songwriters add the phrases they like to a digital sticky note, then copy and paste a desired phrase directly into their song when they're ready to use it. Though these features may indeed help to free users from creative blocks, the song that's written risks resembling a paint-by-number picture. As with a paint-by-number picture, the songs created on Songwriter's Pad are circumscribed by the choices that the designers made when building the app. There may be more original and appropriate phrases that express anger than "you ripped my heart out" and "stormed off fuming," but because they're not part of the app's database, they have less chance of being thought of and less chance of making it into a song composed on this app.

As the Songwriter's Pad example illustrates, there's also less of yourself when you look to apps to supply the grist for your creative acts and encounters. Consider the way many of us now use our search history as a memory prosthesis. Instead of recalling the insights gleaned from an earlier search, we remember only the search terms we used and rely on them to re-create our intellectual path for us.[15] Although this practice is certainly useful and may, at its best, lead us to new insights, it has none of *us* in it. There's no opportunity to take a bite of cake at teatime and, like the narrator in Marcel Proust's novel, launch a meandering journey through our own imagination.

We can see algorithmic lock-in at work in studies that investigate the effects of different media on children's ability to produce imaginative responses. In one study, children in grades one through four were separated randomly into two groups and presented with the same fictional story.[16] One group listened to the story via radio, while the other group watched the story on a television. Afterward, all of the children were asked what they thought would happen next in the story. The researchers rated children's imaginativeness by recording the novel elements (such as characters, setting, dialogue, and feelings) they used in their responses. The children who listened to the radio produced more imaginative responses, whereas the children who watched the television produced more words that repeated the original story. Media scholars have used this study to illustrate the "visualization hypothesis," which posits that children's exposure to ready-made visual images restricts their ability to generate novel images of their own.[17]

Let's revisit our research team's study of teen artwork and fiction in light of the visualization hypothesis. As our focus group participants noted, ready-made visual images are never more than a mouse click or finger swipe away from today's youth. The Internet provides young people access to a greater quantity and wider range of art than in years past. By comparison, youth's access to the literary medium hasn't changed considerably over the years—in fact, the hegemony of the graphic medium may have displaced it. Indeed, linguistic anthropologist Shirley Brice Heath observes that owing to the increase in visual stimuli provided by television and the Internet, youth today are more likely to say "Did you see?" rather than "Did you hear?" or "Did you read?"[18] It is likely that today's young artists draw on this store of visual imagery when they create works of art. Seen in this light, the increased complexity and departure from convention that we detected in teens' art may be less about breaking new ground than about skillfully retreading old. With respect to our analysis of teen fiction, the increased conventionality and use of informal language we observed may reflect the pedestrian language of the tweets, texts, and instant messages that form a substantial portion of youth's daily reading. (We also wonder if these shifts have any relation to the fact that in 2013, after adding a supplemental four hundred word essay to its list of required application materials, Boston College saw a 26 percent decrease in applications. If there were only an app for why one should want to go to BC!)[19] Put succinctly, what seems creative on the surface may actually be re-creative.

In addition to constraining youth's creativity, digital media may also disrupt the mental processes conducive to creative thought. Individuals generate new ideas by reflecting on the world that surrounds them. Reflection requires attention and time (counterintuitive as it may initially seem, boredom has long been a powerful stimulator of the imagination), two things that are hard to come by in today's media-saturated world.[20] Consider the simple act of walking the dog. Before cell phones, it was just you and the dog. In its singular focus, this daily routine (for some, considered a chore) afforded plenty of room for the mind to wander and maybe even stumble on a creative thought. Now, it's just another opportunity to multitask.

In a briefing paper published by the Dana Foundation, cognitive neuroscientist Jordan Grafman expressed the following concerns about our constant state of divided attention: "I think that one of the big trade-offs between multitasking and 'unitasking,' as I call it, is that in multitasking, the opportunity for deeper thinking, for deliberation, or for abstract thinking is much more limited. You have to rely more on surface-level information, and that is not a good recipe for creativity or invention."[21] In support of this claim, there is evidence that individuals who engaged in multitasking displayed cognitive processing that was less flexible and more automatic than subjects who engaged in a single task.[22]

It bears mentioning that breaks in attention can sometimes be good for the creative process, particularly when the goal is to arrive at a sudden insight, or eureka moment. According to the incubation effect, time away from a task enables individuals to restore their cognitive resources, gain new perspective, and avoid impasses.[23] Still, research suggests that it's best when those breaks are chosen by the individual rather than imposed externally in the form of scheduled interruptions.[24] To be sure, today's media landscape provides ample opportunities for self-selected breaks (provided we don't become so absorbed in reading Facebook updates or watching YouTube videos that we abandon our task completely). But this ubiquitous surround also brings frequent interruptions in the form of pinging emails and buzzing phones—or, should these interruptions fail to materialize, even anxiety.

BEYOND APPS

It would be myopic to look at digital media's impact on young people's time and attention without also considering important changes to other aspects of their lives. Indeed, our focus group participants expressed regret that certain changes to the educational environment prevent youth from pursuing their creative interests. In school, arts programming has been sidelined or even eliminated as administrators place test preparation at the center of the curriculum and the heart of the day. One educator bemoaned, "Many of the vehicles that [students] used to be able to express themselves creatively are now gone—theater, arts . . . different electives." This sidelining is most pronounced in struggling schools that typically serve underprivileged youth. At the time of our study, these schools faced the threat of closure if a sufficient percentage of their students failed to meet mandated annual yearly progress goals.

While affluent youth may benefit from greater opportunities for art while in school, participants noted that the regimented quality of their extracurricular activities leaves little room for them to exercise their imaginations outside school. There is little of the precious "time to waste" in youth that is nostalgically recalled by many highly creative artists and scientists.[25] Extracurricular activities have become résumé-building opportunities; students try to distinguish themselves in increasingly impressive ways in the hope of gaining admission to a selective college and, thereafter, to a prestigious internship or job placement. Even the camp experience has been affected, as camp directors feel pressure from parents to provide a documented "value-added" summer experience for their children. As a result, camp has become increasingly structured, the activities more goal-directed. It's hard for imagination to take root, let alone sprout buds, in such arid soil. Indeed, scholars have found that participation in highly structured activities undermines problem finding and creativity.[26]

And what of the workplace? Companies like Google, Facebook, and IDEO claim to value, nurture, and reward creativity in their employees. They go out of their way to create an environment conducive to imaginative thought: innovative office layouts, flexible work schedules—Exhibit A: Google's famous fifth day to pursue a project of one's own design. Such practices would appear to stand in stark contrast to the educational experiences and risk-averse orientations of today's young people. There are two reasons why it does not. Google, Facebook, and IDEO have quite distinct notions of what counts as a good or bad answer to their puzzles. More important, theirs is sanctioned risk-taking. Employees are told, "Here is a context in which you *should* take risks." And, of course, then it is no longer a risk. That said, these programs and recruiter techniques do call for certain creative qualities of mind—ones now described in books that let you figure out whether you "are smart enough to work at Google."[27] They are biased against people who are not good at playing that kind of game—or employing that kind of "app."

WHEN APPS ENHANCE

Although the arguments and evidence presented above concern us, our investigations also give us reason to be optimistic about the creative potentials of apps and other forms of digital media. At the beginning of this chapter, we shared examples of young people—including Molly—using digital media in imaginative ways. These examples illustrate aspects of new media technologies that impress our focus group participants: the lowered bar for entry into creative pursuits, the increased sophistication of what youth can create, and the wider variety of creative opportunities now open to young people. One educator working in a low-income neighborhood reflected: "I would say technology helps. . . . It allows a lot of teens to be creative who wouldn't otherwise: music, robotics, music production." In the words of Seymour Papert, founding member of the MIT Media Lab, and his colleague, Mitch Resnick, head of the Lifelong Kindergarten group at MIT Media Lab, new technologies *lower the floor, raise the ceiling,* and *widen the walls* for youth creators.[28]

Beyond expanding opportunities for creation, there's evidence that certain media activities can enhance individuals' creativity. In one study, researchers investigated the relationship between middle school students' scores on the Torrance Test of Creative Thinking and four types of information technology: computer use, Internet use, video game playing, and cell phone use.[29] The investigators found that all six types of video game playing (including action-adventure, racing/driving, and violent video games) were positively related to creativity. In other words, students who played more video games were more likely to score higher on the TTCT. The researchers found no relationship between students' creativity scores and the amount of time they spent using a computer, the Internet, or their cell phone.

Had the researchers looked at students' specific uses of these technologies, they might have found that certain activities do, in fact, support creativity. That's what researchers in Pamplona, Spain, discovered when they investigated two digital tools designed to stimulate university students' generation of ideas and originality.[30] The first, called Wikideas, uses wiki technology to facilitate the brainstorming process, from generating new ideas to assessing their value. Creativity Connector, the second tool, is a social networking platform that works in conjunction with Wikideas to connect participants and support their collaboration. Study participants were software engineering students enrolled in a project-based learning course. They were instructed to use Wikideas and Creativity Connector to complete a team-based software development project. The researchers found that the tools had a positive effect on the number and originality of students' ideas. The tools also promoted productive, successful collaborations among team members.

In this example, two features of the digital tools are worth noting: *support* and *collaboration*. One of the biggest challenges in the creative process is simply getting started. Wikideas helps students overcome this nontrivial challenge by giving them *support* in the idea-generation phase of creation. Wikideas doesn't come up with ideas for them, just nudges them in the right direction. In this way, it resembles Songwriter's Pad and other apps that attempt to make creating less overwhelming. As composer Igor Stravinsky famously said: "The more art is controlled, limited, worked over, the more it is free. . . . If everything is permissible to me, the best and the worst; if nothing offers me any resistance, then any effort is inconceivable, and I cannot use anything as a basis, and consequently every undertaking becomes futile."[31] And yet, the question remains: Where to draw the line between jump-starting inspiration and locking one in to prepackaged ideas?

As a social networking tool, Creativity Connector is distinguished by its ability to connect people virtually and support their creative *collaboration*. We've already encountered other social networking platforms that perform a similar function, such as Figment, deviantART, and LiveJournal. In his book *Cognitive Surplus*, Clay Shirky celebrates digital media's ability to connect people easily, quickly, and cheaply.[32] Drawing on examples like the Impressionist painters, who lived and worked together in southern France, Shirky argues that collaboration is a central component of creativity. Where collaboration is supported and encouraged—as it surely is online— creativity will thrive.

TAKING STOCK

Just as apps provide new forms of self-exploration and new methods for connecting to other people, they also furnish new means for exercising the imagination. Photo apps allow users to manipulate images in various ways, such as altering color, perspective, and focus. Music apps turn smartphones and tablets into miniature recording studios, while painting apps transform them into easels. Barriers of time, money, and skill are low, expanding which persons can call themselves creators and what they may be able to create. As we've discussed in this chapter, however, the act of creation is circumscribed by the app's underlying code and the developer who wrote it; to paraphrase Lawrence Lessig, the code *determines* the creation.[33] A specific hue of green may not be included in one's painting app; the piccolo might be missing from the music app. Users have little choice but to work within these limitations. The avenues to artistic expression may be many in the app era, but they're often tightly bounded. Creativity scholars sometimes talk about "Big C" and "little c" creativity. The former consists of the truly ground-breaking, original works of art that can change a domain permanently: Stravinsky's *Rite of Spring,* Pablo Picasso's *Desmoiselles d'Avignon,* Martha Graham's *Frontier.* By contrast, "little c" creativity inheres in the realm of daily problem-solving and adaptation to change.[34] Our investigations lead us to conjecture that digital media give rise to—and allow more people to engage in—a "middle c" creativity that is more interesting and impressive than "little c" but—due to built-in software constraints and obstacles to deep engagement—decidedly less ground-breaking than "Big C." These studies also suggest that digital media may have a freeing effect on those young people who already have a disposition to experiment, to imagine, while having a freezing impact on that increasing proportion of youth who would rather follow the line of least resistance.

As we saw in our considerations of identity and intimacy, the digital media do not (at least yet) fully determine how young people think and act. In each case, one can describe scenarios in which the App Generation lapses into a comfortable state of app dependence, as well as a happier scenario in which apps enable youth to have a deeper and more rounded sense of self as well as more fully developed intimate relations with others. With respect to artistic activities, the picture turns out to be even more complex. In the spirit of Marshall McLuhan, we've described how imagination with respect to one medium (graphic expression) is more likely to be enhanced than imagination with respect to another medium (literary expression). When it comes to the matter of creativity, the medium matters. We've noted as well that imagination is likely to be facilitated by the greater ease of communication with others, far as well as near, and by the often powerful vocational and cultural signals in the surrounding community. In our final pages we ponder how these complex factors may be changing the fundamental nature of human society and human consciousness.

SEVEN

Conclusion: Beyond the App Generation

"Civilization advances by extending the
number of important operations which we can
perform without thinking about them."
—*Alfred North Whitehead*

UTOPIAS, DYSTOPIAS

The British writer Anthony Burgess is probably best known for his 1962 novel *A Clockwork Orange,* adapted a decade later (1971) into a memorable movie by director Stanley Kubrick— a work that has in the years since morphed into a cult classic.[1] Briefly, the novel portrays a young ruffian, Alex, who participates all too eagerly in mayhem, rape, and even murder. As Burgess puts it, Alex is generously "endowed, perhaps over-endowed, with three characteristics that we regard as essential attributes of man."[2] To specify: Alex is very articulate; he loves beauty, especially the music of Beethoven; and he revels in violence, specializing in terrorizing urban streets at night.

In an effort to rehabilitate Alex, the state authorizes a form of "aversion therapy." In the course of this two-week regimen, Alex is injected with drugs that compel him to associate violence with extreme nausea. In short order he is transformed into a peaceful if somewhat boring member of the community—in Burgess's phrase, "he is forced to walk a tightrope of imposed 'goodness.'" Reflecting on his book some years later, Burgess put forth his own belief that "it is better to be bad of one's own free will than to be good through scientific brainwashing."[3]

Burgess considered his short novel to be part of the tradition of literary and scientific utopias and dystopias. In Aldous Huxley's *Brave New World,* individuals are deliberately bred—fertilized and conditioned in early life—to become acquiescent members of preordained social classes.[4] In George Orwell's *1984,* the totalitarian state attempts complete brainwashing of its inhabitants, with Winston Smith waging a lonely battle to escape the political clutches of the state.[5] And in a book published at the same time as Kubrick's movie appeared, B. F. Skinner (the poster-child behaviorist- psychologist introduced earlier) described the supposedly benevolent effects of growing up in a society in which human action was completely controlled by a regimen of reinforcement (in lay language, rewards and, more rarely, punishments) of specific actions.[6]

Burgess spurns all such "totalistic" visions, whether they be put forth as utopian or dystopian. As he puts it, "There are few of us who do not reject outright both the Orwellian and the Huxleian nightmares. In a sense we would prefer the repressive society, full of secret police and barbed wire, to the scientifically conditioned one, in which being happy means doing the right thing." Indeed, he says, "enforced conditioning of a mind, however good the social intention, has to be evil."[7]

As a British literary intellectual, critical of technological "fixes," Burgess would likely feel at ease with the world of *L'Éducation sentimentale,* portrayed a century earlier by Gustave Flaubert.[8] While scarcely a backward part of the world—indeed, for many, the Paris of 1850 represented the Apogee of Civilization—there were few signs of advanced technology or media in the milieu of Frédéric Moreau. Frédéric and his circle of friends lived in a world made up of books, pictures, and artistic performances; of stocks, money, and contracts; of gossip, flirtations, and rivalries; of ambitions, achievements, and disappointments. No radio, movies, television, let alone computers, genetic manipulations, pharmacological or electrophysiological conditioning. Flaubert did not engage in a philosophical debate about freedom versus free will, though this is clearly a topic that Frédéric Moreau's circle could have debated eagerly. (As we've noted, they gloried in the ancient art of conversation.) But to many readers, Flaubert's overall message is clear enough: life at twenty is filled with hopes and dreams, while the succeeding years usher in a whittling down of possibilities, regrets of missed opportunities, and poignant memories of things past. Similarly stifling messages permeate Flaubert's other writings, most notably in the portrait of Madame Bovary, the beautiful and zestful provincial wife whose ill-considered liaisons led, seemingly ineluctably, to her sad demise.

But that was France, clearly part of Old Europe. What about the ambience in the wilder parts of the world, specifically the United States as it grew from a garland of colonies in the mid-eighteenth century to a political, economic, and military powerhouse two centuries later?

We receive important clues from the writings of literate Europeans who visited the shores of North America over the centuries.[9] What struck these observers—and we authors seized the opportunity to review their principal writings!—was the pragmatism of Americans; their capacity to put their nose to the grindstone and get something done; their pride in the laws and the political process in their country; in contrast, their ignorance and suspicion of foreign (especially traditional European) society; their discomfort with high art, culture, philosophy; and—it needs to be underscored—their abiding faith in invention, technique, and technology. One encounters little of the air of regret, of roads not taken, that permeates the Flaubertian universe. In their twentieth-century writings, the Canadian Marshall McLuhan and the Frenchman Jacques Ellul did not have to refer specifically to the United States when they alluded to the powers of media and technology, but there is little doubt that they considered the United States a bellwether of what was likely to emerge around the globe. If the world were going to be fundamentally altered by technology, more likely in a dystopian direction, that transformation was likely to take place initially in the United States.

CONCLUSION   159

BEYOND TECHNOLOGICAL DETERMINISM

Anthony Burgess might have been disappointed. In the "app world" described in the preceding chapters there has been no active planning agent—no Mustapha Mond, World Controller in the post–Henry Ford *Brave New World,* no Big Brother of *1984,* no T. E. Frazier, the neo-Thoreauvian architect of Skinner's utopian *Walden Two.* Those scientists and technologists and entrepreneurs who created the hardware and software of the second half of the twentieth century—in the aspiring Silicon Valleys scattered across the earth's surface—cannot reasonably be accused of attempting to fashion, let alone control, all subsequent human behavior. In fact, these digital pioneers were motivated by disparate considerations: sheer scientific curiosity; the search for monetary rewards; the effort to determine the extent to which computers could mimic (or surpass) human intellect in pursuits ranging from the prediction of climate to victory at chess, backgammon, or go; the hope of making human activities easier and more enjoyable to carry out, and, latterly, by the enigma of whether digital ware and neural ware (silicon and synapses) could actually merge. The same mix of motives (commerce, competition, curiosity, consolidation) can be marshaled, more or less, with respect to the intervention of earlier technologies, like the cotton gin or the steam engine, and earlier media, like the telegraph and the radio.

Today we are closer to "being there." It is becoming possible for virtually all of our habits to be initiated and become entrenched, courtesy of our daily (if not moment-to-moment) uses of digital technology; it is becoming possible for us to feel good about this situation. And some of us do. Indeed, the current American interest in "happiness"—and, again, it seems to be an especially American obsession—may reflect a belief that it should be possible all the time to feel positive, to avoid problems, disasters, conflicts, even challenges on which we might fall short.[10] (Note: If only success is possible, such challenges do not live up to their name.)

So how to describe the actual state of affairs? And how do we, the authors—as designated synthesizers—feel about the situation? Without doubt, Technology (the capitalization is deliberate) is a larger part of our lives, from earlier in life, than ever before in human history. The technologies are varied— and this is good—but the strongest influence, particularly among the young, is the pervasiveness of the "app"—the activation of a procedure that allows one to achieve a goal as expeditiously as possible and enjoyably as well. At present, life is certainly more than the sum of apps at our disposal. But the influence of apps is more pervasive and, we believe, potentially more pernicious. And that is because the breadth and the accessibility of apps inculcates an app consciousness, an app worldview: the idea that there are defined ways to achieve whatever we want to achieve, if we are fortunate enough to have the right ensemble of apps, and, at a more macroscopic level, access to the "super-app" for living a certain life, presented to the rest of the world in a certain way. To indulge for a moment in the lowest form of humor: Could just the right ensemble of apps lead to a wholly hAPPy life?

In the preceding chapters, we have spelled out how the app worldview shapes and perhaps constrains the ways in which the chief challenges of youth and early adulthood are negotiated. With respect to identity, there is pressure to present oneself as an impressive, desirable kind of person and to make sure that all signs (and postings) confirm that perhaps precociously crystallized sense of identity. Similarly, with respect to intimacy, the capacity to announce—indeed, to define—one's connections to other persons may preclude fuller exploration, with its heightened vulnerability but also with greater potential for deep and continually evolving relations with truly significant others. Finally, and on a more positive note, with respect to imagination and creativity, digital technologies afford enormous potential for individual or group breakthroughs— provided that the existing apps are treated as approaches to be built upon (allowing us to be app-enabled), rather than ones that constrict or constrain one's means and one's goals (causing us to become app-dependent).

Once again: It's important to bear in mind that our portrait is based on, and applies primarily to middle-class and upper-middle-class youth living in an affluent, developed society. (These represent the same populations that were portrayed by Erik Erikson and David Riesman seventy years ago.) In our research we did not focus on working-class youth, nor on those seen as disadvantaged in terms of economic, social, or demographic variables. That said, and somewhat to our surprise, our informants described much the same story at work in all sectors of society. Teachers described the same tethering to technology, the same hesitation (for the most part) to take risks, the same efforts to create an idealized digital representation of self. Parents of disadvantaged youth were seen as protecting their offspring from challenges or obstacles—and, at considerable sacrifice, making sure that their children had access at all times to smart devices. As anthropologist Shirley Brice Heath has pointed out, if your life has more difficulties, it can either challenge you to create new opportunities for yourself or induce you to revert to quick fixes, be they narcotics or never-ending computer games.[11]

To be sure: Even if our description of today's young people has hit the mark, we can never prove that these features are a direct or even a principal consequence of the pervasiveness of technologies of a certain sort. It is simply impossible to carry out the proper experiment with the needed controls. We cannot divide a state, a nation, or the whole planet into two groups: one given free access to all manners of digital technologies, the other group somehow precluded from any access. (Statements about the effects of any technology—from guns to television—suffer from the same limitation; in a democratic society, one cannot legislate the gold standard of randomized assignment to experimental groups.) And so, in the manner of a persuasive lawyer, indeed a good advocate more generally, the most that we can do is to marshal the relevant arguments, building up the strongest case for the state of affairs that we've observed and the likely reasons for its existence.

One way to think about this conundrum is to imagine whether identity, intimacy, and imagination might have evolved in the manner described, even if computers had ceased to evolve after 1950—the date at which Riesman's and Erikson's pivotal books were published and, as it happens, about the time that Howard entered elementary school. In the sciences, we call this a "thought experiment": no desktops, laptops, tablets, web, Internet, or social networks. Howard can easily imagine such a world, because that's the world into which he (and, for that matter, all previous generations) was born. It's difficult for Katie to imagine, and probably next to impossible for Molly or for Howard's grandchildren to envision—if late, or if lost, what *would* you do without your cell phone?

(Example: At one of the groups on which Howard was permitted to eavesdrop, young people were discussing how they phone family and friends while driving a car or walking their dog. If they could not check in at those times, they lamented, they'd probably never link up with these most valued other persons. Afterward, Howard reminded participants that, when their parents and grandparents were growing up, mobile phones had not yet been invented.)

In the pages of twentieth-century science fiction, we can discern anticipations of the twenty-first-century world—the kinds of worlds envisioned by writers like Isaac Asimov, Ray

Bradbury, Robert Heinlein, Ursula K. Le Guin, or, for that matter, Anthony Burgess. Clearly a digital world could be imagined, even in its actual absence, with the utopian implications that excited, or the totalitarian implications that disturbed, imaginative writers and observers. But could the world that we've discerned actually have come about without technological innovations?

Here is our best guess. Some of the features we've described could have come to pass even if the technology had been frozen at midcentury. To give one example, the reluctance to take risks may emerge because of a belief that there is a best way to do everything and that way is to find the right "app"; but it might also have come about if, for any reason, resources were sharply reduced or competition sharply increased. Getting into a desirable college and securing the right job has long been a goal of young persons (and their parents!)—and understandably so. When opportunities were plentiful, there was a lesser need to walk the straight and narrow, more opportunity to take chances, to forge new paths. (Howard's generation benefited from that brief, Camelot moment.) But when, for any number of reasons, these goals prove far more difficult to attain, a predilection to follow a well-trod path is readily understandable and just as readily justified.

Other features that we've discerned seem much more closely tied to the digital revolution. For example, it is hard to imagine how students could be connected to one another day and night in the absence of mobile phones—and that connectedness has clear implications for intimacy and, as we've suggested, for identity as well. The capacity of human beings to create and promulgate new knowledge is also radically transformed by digital ware. Information that took days or even weeks to track down in Howard's youth—he remembers countless (and often fruitless) treks through the stacks of libraries—can now be located in a manner of seconds on the web; by the same token, findings, claims, and counterclaims are also subject to the 24/7 deluge of information that changes fundamentally the contours, if not the frontiers, of knowledge. We've already seen the differential effect on creativity in two expressive media: literary language and graphic depiction. The pros and cons of "creativity by groups"—be they large "sourced" crowds or small e-salons—have yet to be determined for the digital era.

So far, we've only mentioned the possible causal roles of specific technologies—or of their absence. But of course, many other factors have been at work in the catalysis of generations in earlier times. Clearly, young people will have been defined in part by epochal political and military events—in the United States at the time of the Revolutionary or Civil War, in France or Russia or China at the time of their respective political revolutions, anywhere in battle during the First World War, the Second World War, or the Vietnam or Iraq (or Middle Eastern or Balkan) conflicts. And as we've noted, the consciousness of generations can be engendered by other occurrences, be they financial (the Great Depression, the rise of the consumer society, the eruption of the mortgage crisis), natural disasters (fires, plagues, earthquakes, tsunamis), or manmade occurrences (the *Apollo* mission to the moon, the *Challenger* explosion, the attack on the Twin Towers).

The very identification of other causative factors is salutary and humbling. Even if the current generation is inconceivable in the absence of the technologies of the past half century, these technologies do not and cannot act in isolation. Doubtless, there will be interactions among technological, financial, political, military, natural, and manmade epoch-making events. The most careful students of generational consciousness need and should trace these factors and their interactions scrupulously.[12] (Those who want to understand the American civil rights revolution or the concomitant assertion of women's rights are equally well advised to examine a congeries of events.) And yet we believe that there is a need for some observers to step back and to try to discern the "forest consciousness" that may undergird the many "contributing trees." In introducing and coining the App Generation, that's what we have tried to do.

One more point, which we need to state as clearly and forcibly as possible: Much of what we've written in this book can be seen as critical of the current generation. Characterizations such as "risk-averse," "dependent," "superficial," and "narcissistic" have been asserted, even bandied about. We have to stress, accordingly, that even if these descriptors have merit, in no sense are we *blaming* members of the App Generation. Clearly, these characterizations have come about, at least in significant part, because of the ways in which young persons have been reared (or failed to be reared) by their elders—in this case, Howard's generation and the ones that immediately followed his. If there is a finger to be pointed, it should be aimed at earlier generations and *not* at the adolescents and young adults of our time.

At the head of the chapter, we've affixed a statement by the philosopher Alfred North Whitehead. Though it may be well known among the digerati (Howard first heard it quoted by a leading technologist), we had not encountered it until we were putting the finishing touches on this book. At first blush, the statement sounds just right. One finds oneself nodding in agreement—yes, we value those inventions that allow us to make habitual those thoughts and actions that could assume much time and effort. And indeed, we can think of many manmade devices (ranging from the creation of script to the invention of the credit card) that have allowed us to simplify formerly complex operations and to move on to other things. Is civilization even imaginable without a multitude of labor-saving devices that free our hands and minds? Thank goodness for the "flywheel of civilization"!

Yet on reflection, Whitehead's statement seems increasingly dual-edged to us. For sure, most of us would like to automatize as much as possible—our psychological antagonists— behaviorists and constructivists—would agree. But do we want to automatize *everything?* And *who decides* what is important? And *where do we draw the line* between an operation and the content on which the operation is carried out? The contrasting cases are brought up sharply by Anthony Burgess. The uncivilized Alex decides too much on his own, and that creates mayhem. But the overly civilized Alex has lost altogether the power of decision—all has been molded and modeled by the outside forces. (Remember the end of *The Adventures of Huckleberry Finn:* "But I reckon I got to light out for the Territory ahead of the rest, because Aunt Sally she's going to adopt me and sivilize me and I can't stand it. I been there before.")[13] As we consider the effects of the digital (and, particularly, the app) revolution on our society, we need perennially to ask the question: Do we want to automate the most important operations or do we want to clear the deck so that we can focus, clear-eyed and with full attention, on the most important issues, questions, enigmas?

BEYOND THE THREE IS: THE REALMS OF RELIGION AND ETHICS

As psychologically oriented scholars focused on youth, writing in a post-Riesman, post-Erikson era, we can justify our decision to focus here on the issues of identity, intimacy, and imagination. (Had we studied young children, we might have chosen to talk about trust or initiative or industry; had we focused on older persons, issues of generativity or integrity might have come to the fore.) But especially at a time when claims about life stages and life cycles are being reexamined, we should also touch on a few other spheres in which digital technologies may cast a wide shadow.

First, religion. In one sense, religion (especially as we have known it in the West) can easily be thought of in "app" terms. Many, perhaps most of the rituals involved in regular religious practices can be thought of as "apps"—though of course they are human-initiated and human-choreographed rather than downloaded on one's device. Indeed, the prayer or ritual only works if it is carried out according to the specified procedures. Stepping back, it is also possible to think of the religious life, well lived, or appropriately lived, as a kind of super-app—we must attempt as much as possible to emulate the lives of saints while avoiding the sins (and the sinners) of greed, envy, and other vices.

Yet, perhaps paradoxically, it seems that, in some ways, the app world is antipathetic to religion, or at any rate to traditional organized religion. At least in the United States and much of Europe, young people today are less religious, certainly less formally religious, more skeptical of organized religion, more willing to shift religions, to marry across religious boundaries, and the like. Clearly these trends are not particularly dependent on apps, in the literal sense. Some have unfolded over decades, if not centuries. And yet, the diversity of apps may push us toward defining our own religious practice, in our own way, even our own brand of spirituality, whether or not it happens to conform to that practiced across town or around the neighborhood or even in the next room. And indeed, such exploration can be aided by various apps, which range from Note to God (this app allows users to submit notes to a nondenominational God) to Buddha Box (this app provides chants and sounds to enhance meditation practices).[14] Here as elsewhere, we encounter the lure of app-dependence as well as the option of app-enablement. Ready-made prayer or ritual apps make it easier than ever simply to rely on what the technology affords. The plethora of religion-related apps also makes it possible to choose from a variety of belief systems and practices: in a democratic society, there is little risk of a Big Brother–dictated religious regimen and, instead, much enabling of unusual or even unique theological mixtures. And of course, the apps themselves are only one variable. Users on the adventurous side will concoct their own religious (or atheistic) brew; others will remain ever on the lookout for the one true belief.

Closely related to the arena of religion is that encompassing morality and ethics. Having studied "good work" for many years, members of our research group found it natural to investigate the effect of newly emerging media on venerable, ethically suffused issues such as privacy, protection of intellectual property, trustworthiness, credibility, and citizenship. We did this as part of our Good Play Project.[15] We realized, early on, that aspects of the new media—their speed, their public nature, the ease of accessing, transferring, and transforming information, the possibilities for anonymity or for multiple identities—were creating a virtual Wild West. Ethical issues that, in an earlier time, might have been considered settled were necessarily coming up for reexamination and, perhaps, for reconceptualization.

Our principal findings can be readily summarized. To begin with, as we look across age groups, there is not a radical difference in orientation toward ethical issues. That is to say, we find more similarity than differences across tweens, teens, and adults. Second, there is little evidence in any age group of proactive ethics or exemplary citizenship. When subjects tell us that they avoid missteps, they do so principally out of fear of punishment ("If I send this file illegally, I might get caught and punished"); few state or even imply other, purer ethical motives. The minority who embrace an ethical course is composed primarily of individuals who have themselves seen the harms caused by ethical violations and want to do their part to discourage further ones ("I saw how I felt when someone took credit for lyrics that I had written"). On a more positive note, many young people lament the absence of effective mentors who could model how best to handle an ethical dilemma. Perhaps when such models emerge—and they can come from the ranks of the wise young, as well as the wise old—behavior online may seek and even meet a higher ethical standard.[16]

There may be a more insidious aspect to ethics in the digital era. Even as some individuals believe that ethics should be left to each person (political theorist Alan Wolfe terms this stance "moral freedom"), a surprising number of people assert that ethics is self-evident.[17] We've heard this sentiment frequently from individuals who are part of the Silicon Valley scene or members of groups that promote digital freedom. In the spirit of Google's motto, "Don't be evil," there is the apparent belief that people of goodwill can be counted on to behave in a righteous way. What we know about human behavior is that it is all too easy for individuals to *believe* that they have good motives and behave well, even though informed observers dispute this characterization.[18] It is also easy to believe that others of goodwill necessarily concur with one's own views. ("It's obvious that we need to protect individual privacy" versus "What people *really* want is complete transparency about all matters.") Being genuinely ethical requires much soul-searching, conversing with informed peers, a willingness to admit that one has been wrong, and striving to do better the next time. These steps are far more difficult to execute than a simple delineation of what is ethical and what is not. ("It's OK to mislead a novice in World of Warcraft, because, after all, it's only a game.") Put differently, apps may help to raise consciousness about ethical conundrums, but they cannot confidently designate the best course of action in a particular situation. Take note, Professor Whitehead!

Building on our several years of research, we have in fact initiated efforts to help strengthen the ethical muscles of those involved in the digital world (which includes just about everyone). One such effort, in collaboration with Project New Media Literacies, involved the creation of an ethics casebook, called *Our Space,* for use in secondary schools. Another, in collaboration with Common Sense Media, entails the creation of a guide to digital citizenship for use in both middle and secondary schools.[19]

In such efforts, we make no claims that we have clear-cut answers to vexing issues of privacy or intellectual property or the meaning of membership in a virtual community. The issues are too novel and the terrain changes too quickly. Instead, our efforts involve the posing of enigmatic problems and the engagement of young people in discussions of what they might do in particular situations and what the consequences might be. So, for example, young people discuss what to do when a girl posts damaging information about her family on her Facebook page, or a boy circulates lyrics written by someone else without any attribution; or when someone transmits a photo of an athlete engaged in impermissible behavior on the night before a big game. Guidance for the discussion typically includes citing of existing laws and regulations, generally accepted practices, and possible penalties as well as promising models. Often teachers, parents, and other elders benefit as much from these discussions as do the young persons for whom they have ostensibly been designed.[20] Importantly, we have here an area of contemporary life in which individuals of different ages, backgrounds, and sensibilities can educate one another about possible best practices.

EDUCATION IN THE ERA OF THE APPS

Which leaves for consideration what may well be the most important issue: how the digital media are affecting and may continue to affect education. We begin with a vital implication that has yet to be fully acknowledged: education is no longer restricted to K–12 or even K-through-graduate-school.

It is lifelong! Education (and, it must be added, miseducation) begins as early as the time when toddlers can play with phones, tablets, or remote control devices, and it continues as long as individuals wish to be involved actively in the world. (Acknowledging this reality, Howard has lobbied to change the name of the school at which he teaches from the Harvard Graduate School of Education to the Harvard Graduate School of Lifelong Learning.)

Digital devices make possible a degree of individuation and pluralization that would have been virtually (excuse the pun!) inconceivable in earlier epochs.[21] We live in an era when individuals can study, or attempt to acquire a skill, when they want to, at a pace of their own selection, alone or with others, with or without badges or other forms of certification—no two persons have to be educated or to educate themselves in a single mandated way. One-size-fits-all curricula and pedagogy deserve to be anachronistic, if not indictable offenses. The possibility of entering and mastering important topics and skills in multiple ways is also enabled by the digital media. There are now many ways—involving many media and varying in degrees of proactivity—in which to learn to play chess or the piano, to speak French or read Chinese characters, or to gain knowledge of economics, statistics, history, or philosophy. Furthermore, in our era, digital devices also enable a degree of collaboration with those far away, as well as those nearby, which would not have been possible or even conceivable in earlier eras. This is all to the good!

But less palatable aspects of learning also mark a digital era.

One is the threat to residential learning at college. To be sure, residential learning is expensive, and its dividends are not always immediately demonstrable. Why pay thousands of dollars and move to another city, if one can sit at home and master a well-designed MOOC (massive open online course)? But there are many reasons for cohorts of learners—whether in liberal arts colleges or in professional schools dedicated to law or medicine or nursing or engineering—to spend time together, in the company of well-trained and informed teachers and mentors. So much of what is important in work is not easily, or not usually, put into words; it is best picked up by being around those who carry out key practices in well-worked-out ways every day. Sixty years ago, philosopher Michael Polanyi pointed out that one could read about science for one's whole life in a far corner of the world; but this literary immersion would not compare with spending a few weeks in a well-run scientific laboratory in the developed world.[22] We might well ponder whether we would want to have surgery performed or our bridges built or our case presented to a jury by an individual who may have received a high score on a certification exam but has never stood shoulder to shoulder with peers and mentors in an actual work setting.

Should an app mentality be imposed on lifelong education, an even greater danger lurks. All over the world, prodded by a consensus among Anglo-American policymakers, there is a belief that there is one body of knowledge that deserves to be mastered (typically, the STEM tetrad of science, technology, engineering, and mathematics—itself an "app quartet"); a best way to present it; and a best way to measure it (typically by a multiple-choice machine-administered and machine-scored examination issued by the Educational Testing Service). And there is as well the dream (or is it a nightmare?) that one can array all students, all teachers, indeed all nations, in terms of their performances on these allegedly fair and comprehensive instruments. Almost none of the highly creative individuals of the past that Howard has studied—among them painter Pablo Picasso, poet T. S. Eliot, dancer and choreographer Martha Graham, leader Mahatma Gandhi—would have stood out on such measures.[23] And among contemporary artists, it can be said that Princeton University would have been poorer without painter Frank Stella, just as Harvard University would have been the loser had cellist Yo-Yo Ma or poet John Ashbery or actor John Lithgow not elected to study there. (We hope that they also appreciated their broad liberal arts education.)

We have no doubt that, on the part of some, the motivation to carry out "objective education with objective measurements" is laudable. And we have no doubt that some individuals have misused a system in which more subjective or idiosyncratic or hyper-pluralistic approaches were sanctioned. But we are equally convinced that education is too important, and too subtle, to be outsourced to the Educational Testing Service, or to what Finnish educator Pasi Sahlberg wryly terms the GERM approach of Anglo-American education: Global Educational Reform Movement.[24] Using approaches to health care as a model, Howard has commented, "When it comes to health care, there's a lot to be said for the 'checklist approach' favored by surgeon Atul Gawande. But when it comes to education, this sector remains in many ways an art, and one does well to follow the advice of surgeon Jerome Groopman—listen, listen hard, and then listen even harder."[25] Contemplating the current Anglo-American intoxication with objective measurement of certain performances and a concomitant insensitivity to differences in human gifts and human aspirations, we have been concerned about an approach to education that is overly reliant on apps.

In fact, we've faced this conundrum of assessment with our own work. In studying GoodWork we have tried to define the features of such work with clarity.[26] A group of teacher colleagues in India converted our prose definitions into carefully calibrated ten-point scales of ethical behavior. At first glance, this feat was an achievement that sharpened our thinking. And yet, it seemed to imply a precision in assessing ethics that could not realistically be achieved. Asked what he thought of the scoring system, Howard praised the devisers for their thoughtfulness and their diligence. But he added that perhaps the system promised more than it could deliver. Instead, Howard suggested, "Why not simply indicate where the school is headed in the ethical sphere—an arrow pointing 'up' means that progress has been made, while one pointed downward suggests the need for more work?"

Leading a series of conversations in which college freshmen reflected on their lives, Howard asked the dozen or so students to indicate their personal goals for the series.[27] He was surprised by one of the responses: "I don't want to work on issues where there are no answers." Howard made a mental note of this response but at the time said nothing to the student. Later, after the sessions were over, Howard spent some time with the student and learned that he intended to major in biology (he wanted to become a surgeon) but also in philosophy. Since philosophy has traditionally focused on questions for which there are no answers, or at least no glib or definitive ones, Howard asked the student why he had said that he did not want to spend time on questions with no answers. The student said, "I don't like sessions where people just talk around in circles." But he then admitted, what seemed evident to Howard, that an interest in philosophy is hard to square with a belief that all questions must have neat answers. We suspect that this eighteen-year-old, growing up in an "app world," was impatient with conversation that did not seem goal directed. And this sentiment, which seems to be widespread, spells trouble for the study of the traditional liberal arts: interests in literature, philosophy, and history are difficult to sustain if you believe that all knowledge is—or should be—susceptible to an algorithmic process culminating in a consensually accepted correct answer or "product."

In fact, the two strands in the student's psyche epitomize well the enigma entailed in the epigraph to this chapter. On one reading of Whitehead's words, the avid student is well advised to be able to automate as many features of living as possible: whether it entails mastering the human anatomy so that surgery can be performed expertly or avoiding conversations that lead nowhere and seem to be time wasters. And yet, how can one know in advance *which* circumstance in surgery might require an instant decision involving an obscure bit of anatomical knowledge or *which* stray comment in an evening bull session might cause one to rethink an important life decision "just in time"?

There's a curious disjunction today in the world of those who speak publicly about educational means and goals. On the one hand, particularly among leaders in business, there is much talk about twenty-first-century skills—the "four Cs" of critical thinking, creative thinking, collaboration, and community.[28] On the other hand, almost all educators (or, perhaps more accurately, educationists) in positions of authority in the United States call for the kind of constrained curriculum and traditional standard tests that at their best capture skills of a bygone era. Given this disjunction, the status of digital learning in general, and apps in particular, is invoked by both sides in the debate. Those favoring the more open-ended skills focus on the enabling qualities of the digital world, whereas those defending the traditional skills seek to mobilize digital media to increase the efficiency and effectiveness of existing delivery and assessment processes.

Let's dive directly into the world of educational apps. Our survey suggests that the majority—one might even say, the vast majority—of educational apps encourage pursuit of the goals and means of traditional education by digital means. They constitute convenient, neat, sometimes even seductive pathways to accomplish what were already goals in an earlier era: mastering concepts, learning arithmetical operations, identifying geographical locations or historical figures or key biological or chemical or physical processes. We could dub them "digital textbooks" or "lectures" or "preprogrammed educational conversations." Decades ago major behaviorist B. F. Skinner called for teaching machines that would automate the traditional classroom, allow students to proceed at their own rate, provide positive feedback on correct answers, and either repeat a missed item or present that item via another pathway.[29] Those sympathetic to Skinner's brand of psychology and to its associated educational regimen would easily recognize many apps today and would likely nod in approval at their slick, seductive interfaces.

Just as generals are prone to fight the last war, it is probably not surprising that the first generation of educational apps resemble pre-app education. (In fact, no less an authority than Marshall McLuhan noted that new media always begin by presenting the contents of the previous media.) Yet in our view, this tried-and-true pathway represents a missed opportunity. (And given how slowly change happens in our public education system, it risks becoming codified in the curriculum for many years to come.)

Let's turn the educational challenge on its head. What features are *newly* enabled by the new media, and how can one create and deploy apps that take maximum advantage of these affordances?

As we see it, the new media offer two dramatically fresh opportunities. One is the *chance to initiate and fashion one's own products*. As we transition from web 1.0 to web 2.0 and beyond, there is no reason anymore simply to respond to stimuli fashioned by others, no matter how scintillating and inviting they may be. Rather, any person in possession of a smart device can begin to sketch, publish, take notes, network, create works of reflection, art, science—in short, each person can be his or her own creator of knowledge.

The second opportunity entails *the capacity to make use of diverse forms of understanding, knowing, expressing, and critiquing*—in terms that Howard has made familiar, our multiple forms of intelligence. Until recently, education was strongly constrained to highlight two forms of human intelligence: linguistic and logical-mathematical. (Indeed, until the end of the nineteenth century, linguistic intelligence was prioritized; in the twentieth century, logical-mathematical intelligence gained equal if not greater importance.) The digital media enable a far greater spectrum of intellectual tools. Not only does this opening up of options allow many more forms of expression and understanding. It also exposes young people to different forms and formulations of knowledge. It gives additional forms of expression to all, and most especially to those whose strengths may not lie in the traditional arenas of language and logic—for example, to future architects, musicians, designers, craftspeople, and maybe even creators of innovative new software.

High time for an example. We turn here to *Scratch,* a wonderful application created over the past two decades by Mitch Resnick, a valued colleague at MIT, and his colleagues. Building on Seymour Papert's pioneering work with LOGO—a prototypical example of constructivist education—Scratch is a simple programming language accessible even to youngsters who have just reached school age. By piecing together forms that resemble pieces of a jigsaw puzzle, users of Scratch can create their own messages, be these stories, works of art, games, musical compositions, dances, or animated cartoons—indeed, just about any form in any kind of format. Moreover, users of Scratch can and do post their creations. Others around the world can visit these creations, react to them, build on them, and perhaps even re-create them in their own favored symbolic system.

The genius of Scratch is twofold. First of all, it opens up a plethora of modes of expression, so that nearly every child can find an approach that is congenial with his or her own goals, strengths, and imaginations. Second, educational ends and priorities are not dictated from on high; rather, they can and do emerge from the child's own explorations of the Scratch universe. In that sense, Scratch brings pleasure and comfort to those who believe in the constructivist view of knowledge. Not only are users building their own forms of meaning and constructing knowledge that they personally value, but they are epitomizing the claim of cognitivists that one learns by taking the initiative, making one's own often instructive mistakes along the way, and then, on the basis of feedback from self and others, altering course and moving ahead.

Still, just as a hammer in the hands of a vandal can be used simply to strike every item in sight, it would be possible to misuse Scratch, to miss its genius and to convert it into yet another behavioral tool. This less happy outcome occurs when adults—no doubt, well meaning in most cases—"hijack" Scratch in the exclusive service of traditional educational goals and means. For example, in an educational setting wedded to a behaviorist approach, it would be possible to use Scratch to model one specific way of drawing objects in the world or for providing the definitive model of how to represent fractions or write a sentence, a paragraph, or indeed an essay.

We see, then, that the app itself is never a foolproof avenue to one or another educational use or philosophy. Depending on the context in which it is used, and the priorities of the educators (which includes those present in the classroom, lurking at home, or at their drawing boards or computer screens at an educational publisher), one can skew the same application toward app-dependent or app-enabling ends.

Nonetheless, we have no intention of letting the app-creators off the hook. Those who design apps can skew them toward *dependence;* this is what happens when powerful instructions and constraints are built into the app. Recall our discussion in the previous chapter of Songwriter's Pad, an app for writing songs and poems on the iPad. Choose a mood from the list of available moods, and the app returns a corresponding list of words and phrases associated with that mood that you can then insert into your song or poem. We don't doubt that some people will use this app in creative, unexpected ways. However, the constraints built into Songwriter's Pad—in the form of packaged "bites" of poetic words and phrases—strike us as leaning toward app-dependence. Alternatively, app designers can skew an app's constraints toward *enabling;* this is what happens when, à la Scratch, the apps are wide open, when they offer multiple forms of expression, and when the responses from adults and other users are not constrained.

Nor do we intend to leave adults—be they parents or teachers—off the hook. Depending on the milieu at home or at school, adults can either signal that apps are simply the latest and most efficient means to a given educational goal—typically, the traditional "mastery of prior knowledge" that has been the staple of education for many years. Or they can signal that apps represent a new avenue for individuals to explore different pathways, to record their own forms of understanding, and to solicit reactions from others, ranging from those with much knowledge to those who may themselves be edified by the product or project in question.

Take, for example, a new app released in the summer of 2013 by Sesame Workshop, famous for the innovative television series *Sesame Street.* According to its creators, the Big Bird's Words app lays the groundwork for learning new words. Using text recognition technology, the app prompts children to identify various words—grouped in categories—in their surrounding environment. The online demo shows a young boy of three or four working in the food category. He chooses the word *milk* from a list of food words (each item in the list has a picture next to the word), then holds up his smartphone to a milk carton. Big Bird says, "Milk," and congratulates the child for finding the correct word.

Used in an enabling spirit, this app can encourage children to explore the words around them and connect these words to their daily activities. This might lead to exploration of other words in the children's environment but perhaps not in the app's lexical database. These explorations might even involve discussions with parents and siblings. Used in a dependent spirit, however, the app might engender an overreliance on it for word recognition and perhaps send the message to some children that the only words worth knowing are those that are included in the app's database. Seen in this light, the app-dependent use limits how children explore and learn from their world.

And so we look toward mindful adults—whether new young parents or wise elderly trustees—to furnish the settings within which apps will be encountered and used. It's in our hands to provide nudges in the direction of flexible use of apps; to offer initial scaffolds in the form or use of apps but then to remove these as soon as feasible; and to sanction the implementation of spaces and of times in which one puts aside the devices and the apps and fends for oneself. Seth Kugel, who writes the "Frugal Traveler" column for the *New York Times,* describes the freedom encountered when he renounces his dependence on travel apps: "I believe everyone should use the vast online database of the travel world with moderation. Save a day or two for spontaneity: seek advice from a stranger on the Seoul subway; take a day to explore an Italian town just because you stopped there for gas; trust your instinct to find a Parisian bistro to call your own. Maybe you'll find out later that its croque-madame has been praised 717 times on TripAdvisor. Who cares? You discovered it yourself."[30]

When we began to write this book, neither of us had in mind the educational writings of Alfred North Whitehead, to whom we owe the tantalizing epigraph at the head of this chapter. Yet as it happens, we find extremely useful Whitehead's own approach to education, as expressed in his little volume *The Aims of Education*.[31] In surveying the steps involved in becoming an educated human being, Whitehead identified the recurring sequence of romance, precision, and generalization.

As Whitehead saw it, genuine learning begins when one is excited, moved, inspired, or stimulated by an early encounter with a question, phenomenon, or mystery—this is the time of romance. But one remains stuck at this point, or becomes bored or alienated or anxious, unless one can begin to acquire tools that allow one to gain a firmer understanding of the initially seductive phenomenon. (Of course, the acquisition of precision can be done in many ways, ranging from the strict behaviorist regimen to the flexible, exploring constructivist tack.) Ultimately, the acquired knowledge and skills need to be put into a broader context; related to other forms of knowledge and understanding; and serving as a prod to further learning, with its initial romantic encounters.

Please note that by no means are we dismissing the importance of learning what prior generations have already established. We do *not* believe that individuals can or should construct all of knowledge on their own. That would be absurd. Indeed, new knowledge must be built on what has already been consolidated by earlier thoughtful individuals and groups—in Matthew Arnold's well-turned phrase, "to make the best that has been thought and known in the world current everywhere."[32]

Our point is different. Put directly, we are not unduly worried about avenues to precision: many exist. What we are here urging is that apps can and should facilitate the initial romance; present multiple ways of attaining precision; and, in the end, provide ample opportunities to make novel as well as expected use of what has been learned. This stance should occur both with respect to constrained educational goals— say, the understanding of multiplication—and with respect to the broadest educational goals—say, the appreciation of how scientific knowledge is created and used and misused. Indeed—and here is where we draw the line sharply between behaviorists and constructivists—precision should always be the *means* toward making knowledge one's own and using it ultimately to raise new questions and build additional knowledge.

You may well be saying, "You authors are certainly giving apps a hard time." And we might even plead nolo contendere— that is, we wouldn't dispute your characterization in court. Time to say, loudly and clearly, that there are many wonderful apps, designed to do well, better than most of us could do on our own, what needs to be done. To paraphrase Whitehead, they free us to focus on what we want to do or what remains to be done. Moreover, many apps have been created by ordinary citizens who have discerned a problem and have found a way to address it, to fix it. Two cheers for apps!

APPS FOR A BETTER WORLD

An impressive example of what apps can accomplish comes from the work of an organization called Code for America. As explained by founder Jennifer Pahlka, Code for America fellows are chosen to undertake a year's assignment.[33] During that period they work closely with public officials in city governments to create apps that solve problems identified by administrators or citizens. These needs range from finding the optimal flow of traffic to the placing of children in appropriate schools and to helping people who use food stamps to locate high-quality, affordable foods. To give an example, an app developed in Boston to identify potholes is open source and can be used by any other municipality.[34]

What's striking about Code for America is that its fellows can often solve problems at a fraction of the estimated cost, and a fraction of the estimated time span, than anyone at city hall could have anticipated. A belief in the power of apps, coupled with a sense of important problems and how they might be addressed efficiently and effectively, yields a win all around. And of course, the existence of Code for America does not preclude the addressing of problems that are more vexed and do not lend themselves to a neat application. Indeed, in the ideal, it can free officials to devote more time to larger, less tractable challenges.

For those of us in the social sciences, there is an "apt" analogy. Forty-five years ago, at the same time that Howard was restricted to a few media outlets, he also had to perform most statistical tests with pen and pencil or the aid of a handheld calculator. These were time-consuming tasks. But in carrying out these computations, Howard got to know his data very well. Nowadays, powerful computers (along with more sophisticated statistical techniques) allow one to arrive at findings at warp speed. If the time saved gets translated into closer scrutiny of the data and a deeper, more cogent analysis of what they mean, the apps have been invaluable. If, however, they create the illusion that the data (let alone the "big data") speak for themselves or simply make the researcher impatient to collect and post the next trove of data, then the app has not been so helpful.

As an example from a very different realm, consider the creation by composer Tod Machover of a work called *A Toronto Symphony*. Dubbed "America's most wired composer," Machover has pioneered the use of electronic and digital instrumentation in many compositions; he has also created new approaches to scoring and devised toylike instruments that can be played by individuals with no formal training in music.

A Toronto Symphony breaks new ground. This massively collaborative composition (which goes beyond crowdsourcing) involves ordinary citizens—chiefly from Toronto, though anyone can play—in the co-creation of a major symphonic work. For one section of the work, inhabitants are invited to record and submit sounds that they find expressive of the city. (This can be seen as a contemporary version of the street sounds in the opening section of George Gershwin's *An American in Paris*.) For other sections of the symphony, Machover and his team have created apps that, like paintbrushes, can be applied in different colors and with differing intensity. These apps allow users to shape a melody sketched by Machover, both in terms of its broad contour and its finer details, or to create their own collages and mashups of musical material from the piece. Sampling and studying these various contributions garnered over several months, Machover becomes the final creator of the work. But as he has put it, "If it feels in the end like basically my piece no matter what, or like a mash-up of other people's stuff that I facilitated, I think that would be less satisfying . . . but if it's something that couldn't have been made without each other, it will feel really good."[35]

Machover's symphonic composition differs in its goals and methods from Pahlka's Code for America. Whereas Pahlka is trying to solve vexing urban problems, Machover is creating a tribute to an admired urban environment: engineering versus art. But note that creating a musical work in the digital landscape is also a feat of engineering, while creating an effective municipal app is also an artistic feat. Probing further, we see the contributions of ordinary, nonexperts ("what used to be called the audience," as one pundit has put it): in the case of Code for America, suggesting problems that need to be tackled and using the solutions created by the fellows. In parallel fashion, ordinary nonexperts evince their best efforts to orchestrate sections of a work, and once the work has been completed and performed, the audience members can assess the success of the piece. In the finest sense, we see at work joint efforts between citizens and experts, and a fine balance between algorithms (apps) and taste (app transcendence).

LOOKING AHEAD

On one reading, it may seem that we see apps and the App Generation as moving inexorably in the direction of ready-made solutions to existing problems. In this unappealing scenario, identities will be more superficial, packaged less interestingly, idiosyncratically, less meaningfully consolidated; intimacy—even if it proves more robust than privacy—will be more superficial, more tenuous, less likely to evolve over time; and imagination will be enhanced chiefly for evident problems with evident routes toward their solution. Or, extending beyond our individual young subjects, it may seem that, in spheres ranging from religion to education, the plurality of apps, and the uses to which they're currently put, lean strongly in the direction of dependence, not enablement.

But the App Generation (and its successors) need not accept these trends. As individuals, as groups, as cultures, people can decide at certain times, or under certain circumstances, to *dis*engage from the digital world, to explore paths on their own, to form identities, achieve degrees and forms of intimacy, and forge creative directions that had never been anticipated before. (Of course, as Jacques Ellul might have quipped, disconnecting technologically may prove easier than challenging the consciousness created by technology.) The birth of writing did not destroy human memory, though it probably brought to the fore different forms of memory for different purposes. The birth of printing did not destroy beautifully wrought graphic works, nor did it undermine all hierarchically organized religions. And the birth of apps need not destroy the human capacities to generate new issues and new solutions, and to approach them with the aid of technology when helpful, and otherwise to rely on one's wit.

Of course, other potent factors are at work. In this book, we have not spoken much about the ambition and reach of vast multinational corporations or of totalitarian states.[36] For every major medium of communication that began as the product of human imagination, one can tell a story of how megacorporations eventually came to dominate the media and to determine how human beings interacted with them. Google, Apple, Amazon, and their less prominent peers have tremendous power and access to data of a size and scale that not even the most imaginative science fiction writers— H. G. Wells, Jules Verne—could have anticipated a century ago. It would be a brave person who would predict that the fate of corporation-devised and -sold apps would be different; and it would be a naive person who would simply assume that such power will inevitably be dedicated to benign uses.

We must also acknowledge the possibility of powers even greater than those associated with megacorporations and powerful political entities. As we come to understand better our genetic and neurological nature, there will be attempts to reconfigure our species, more or less aggressively, and to usher in a so-called *singularity,* in which the lines between computer and brain, machine and human, mortality and immortality become blurred or blended or disappear altogether.[37] As more than one wag has put it, "The question is no longer, 'Are computers like us?' but rather, 'Are we like computers?'" To the extent that these impulses are realized, human tendencies to resist or transcend apps will evaporate. Just as surely as the reach of Big Brother in *1984* or the programming of Alex's brain in *Clockwork Orange,* apps will come to control our lives.

And so we come full circle to the questions raised by Anthony Burgess. Is it better for our species to tolerate our imperfections—our individualized identities, our idiosyncratic forms of intimacy, our stumbling but earnest and perhaps unique efforts to be creative? Or should we attempt to uncover, or create, the full spectrum of apps or the super-app needed or wanted so that you—we—can pursue a certain view of the Good Life? One does not have to embrace a romantic notion of free will to acknowledge that this is a genuine choice, and one that future (if not present) generations will need to make—individually and perhaps collectively.

In closing, let's revisit our central metaphor. As a species, we face a choice. Apps are not going to disappear, and there is no reason why they should. The question is whether we are going to become increasingly app-dependent—looking for an app in every situation and spurning any that lack a ready app? Or will we become app-enabled—using established and new apps to broaden our repertoire of possibilities? Or even, on rare occasions, tossing technology to the winds, app-transcendent? Perhaps, in the spirit of an analog (rather than a digital) age, and recalling the ticking clock featured on the front page of the *Bulletin of the Atomic Scientists,* we should monitor whether, over time, the arrow points toward greater dependency (which for us would be dystopian) or toward greater enabling (which for us would be utopian).

As this book was just about finished, Howard had the opportunity to talk with his grandson Oscar, then aged six and a half, about his experiences with digital media. Except for Howard's checking with Oscar's parents beforehand, Oscar was not prepared or prompted in any way for the chat. He allowed Howard to record the conversation and, in fact, when the interview was concluded, showed Howard how to shut off the recording function on his iPhone.

Not surprisingly, as a child born in 2005, Oscar has always been surrounded by digital media. He is completely conversant and comfortable with the terminology and jargon. Howard asked him what would happen if "Opa" (as Oscar calls his grandfather) took away his iPhone.

> OSCAR: I would not be sad, I still have a computer.
> HOWARD: Oh, what's it like?

OSCAR: Bigger than my mom's.

HOWARD: What do you do on it?

OSCAR: Search on toys, go to dot com, to do something like herofactory dot com. Little things. . . . I can write a little code into the line, so I can play some sort of game.

Howard was a bit taken aback at Oscar's ease with terminology (dot com) and activity (write a little code). And so Howard asked him if he ever "Googled" anything. The following exchange ensued:

OSCAR: I Google *everything*, Amazon, like anything I need to go to Google or write it down.

HOWARD: You sound a bit exasperated.

OSCAR: Kind of, but I['m] not sure I know what "exasperated" means.

Howard moved next to what one does and what one does not do with computers. Here Oscar made a very clear distinction:

HOWARD: I grew up without computers. What do you think that was like?

OSCAR: People would do all chores and more chores and more chores and no fun.

HOWARD: No fun?

OSCAR: A little bit, but not much fun.

HOWARD: Do you use computers for school and study?

OSCAR: I don't really do [those] things. I just use my computers for fun.

HOWARD: How do your mom and dad use computers?

OSCAR: For only one thing . . . work. My mom downloads things that she has to do, like, does work about food in my school [Oscar's mother is doing graduate work in food science].

It appears, then, that Oscar makes a rather sharp division: kids/computers/fun versus adults/no computer/no fun or adults/computers/work.

But were computers merely a source of pleasure and amusement? Howard decided to push Oscar a bit on what digital media did and did not mean to him, and what they enabled or prevented. This conversation proved most illuminating with respect to how Oscar sees the world—his digital worldview:

HOWARD: How do you feel when your parents say, "Put it away"?

OSCAR: Feel a little blue, a little blue [said in a slightly plaintive tone].

HOWARD: How would you feel if your parents took all your computers and phones away for a few weeks?

OSCAR: I'd feel a little blue, but I could actually have a little more freedom . . . play with my toys, play with Aggie [his then-eight-month-old sister], go to places with Mom and Dad.

HOWARD: What do you mean by "freedom"?

OSCAR: Mostly people have technology [his word,

no prompting from his grandfather], they are watching every game, and [makes a boring sound] and do it all day, and [don't] do anything else, but just watch TV. . . . So you can play with toys and things like that.

Oscar is certainly not a student of digital media, nor has he read about utopias and dystopias. Nor have his parents or grandparents discussed with him the ambiguous seductions of the digital media. And yet, at the tender age of six, he already senses that one can become a prisoner of the new technologies and that a world beyond them is beckoning to be explored . . . there were but time and space to do so. He does not need to be a participant in the toy-playing experiment carried out by Elizabeth Bonawitz and her colleagues. In some sense he has attained the insight embedded in that study: even though a well-demonstrated toy or well-designed app has its virtues, there is also virtue—and even reward—in figuring out things for yourself on your own time, in your own way.

With essayist Christine Rosen, we worry about the "ultimate efficiency—having one's needs and desires foreseen and the vicissitudes of future possible experiences controlled." With poet Allen Tate, we spurn a world in which "we no longer ask 'is it right,' we ask 'does it work?'"[38]

As authors, we get the privilege of last words. For ourselves, and for those who come after us as well, we desire a world where all human beings have a chance to create their own answers, indeed, to raise their own questions, and to approach them in ways that are their own.

Methodological Appendix

In 2008, we conducted interviews with forty long-standing classroom teachers (twenty-four men, sixteen women) to cull their observations about how current students may be different from the students they taught in the predigital era. Educators and researchers affiliated with Harvard Project Zero recommended these educators to us based on their years of experience and teaching excellence. Participants averaged 23.5 years of teaching experience, and all but two had been teaching since 1992.

Participants were drawn from eighteen schools in the greater Boston area and one in central New Hampshire. All served students from affluent families. In total, we interviewed educators from two middle schools, two colleges, and fifteen high schools.

The educators in our sample represent a broad range of intellectual disciplines, including history (6), general social studies (1), English and/or English literature (6), foreign language (2), art (5), theater arts (7), music (5), biology (3), chemistry (1), physics (2), athletics (2), and general education (1). Several also coached a sport or, in the case of boarding school teachers, served as housemasters, and could therefore comment on their students' lives outside the classroom.

Two researchers conducted the interviews, which followed a semistructured interview protocol and lasted approximately 90 to 120 minutes. Participants were first asked general questions about the changes they've noticed in various aspects of students' lives, including academic engagement and performance, peer relationships, and extracurricular activities and interests. In an effort to avoid biased responses, we deliberately did not raise the topic of digital media during this part of the interview. However, participants typically introduced the topic on their own, and we were ready with a series of follow-up questions.

All but one of the interviews was audio-recorded. Both interviewers took detailed notes during each interview, which they later synthesized into a single record. Attached to each record was a briefer topsheet that summarized key points of interest. Halfway through the interview process, project staff constructed a matrix grid to organize salient data from each master record and topsheet. Categories were determined by the strength and frequency of a finding and later amended as more data were captured.

II. FOCUS GROUPS

Members of our research team conducted seven focus groups between May 2009 and March 2011. Participants were fifty-eight veteran professionals who each had over twenty years of experience working with young people (roughly ages twelve to twenty-two) in a variety of settings. The professionals included psychoanalysts; psychologists and other mental health workers;

camp directors and longtime counselors; religious leaders; arts educators; and high school teachers and after-school educators who worked primarily with students living in low-income neighborhoods.

The focus group facilitator asked participants to reflect on changes they have observed in youth over the last twenty years and to offer their thoughts on the causes of these changes. Each participant was given 5 to 10 minutes to share his or her initial reflections, with the facilitator asking clarifying questions and summarizing themes as appropriate.

Members of the research team followed up with questions that encouraged participants to elaborate on their answers and invited them to respond to comments made by participants in earlier focus groups. The majority of these follow-up questions related to the "three Is" (identity, intimacy, imagination) identified as dominant themes from earlier, one-on-one interviews with veteran educators. To guard against leading questions, digital media were not introduced as a topic of conversation until a participant explicitly made reference to them.

Following each focus group, researchers compiled their individual field notes into a single memo summarizing the major themes discussed. One researcher then synthesized these themes into a series of formal reports, one for each group of professionals.

III. TEENS' CREATIVE PRODUCTIONS

Between February 2011 and August 2012, we conducted three related studies of artwork and fiction writing produced by teens between 1990 and 2011. We analyzed 354 works of high school students' visual art, 50 short stories written by high school students, and 44 short stories written by middle school students.

Visual Art

Our visual art sample included a random selection of 354 artworks published in a teen literary and art magazine based in Massachusetts. The same husband-and-wife editorial team has published the magazine, called *Teen Ink*, since its inception in 1989. Half of the pieces in our sample (n = 177) appeared in issues published in 1990–1995, and half of the pieces appeared in annual issues published in 2006–2011. Though most submissions are two-dimensional, a small number of pieces are photographs of three-dimensional artworks, such as sculptures and installation work.

We selected artwork to include in our sample from the "Art Gallery" page that appears in every issue of *Teen Ink*. We used a random number generator to select three pieces from each issue, with the exception of the February 1991 issue, which was not available in the *Teen Ink* archives. To balance out the number of issues included in the early and late periods, we additionally omitted the December 2011 issue.

The low print quality of the early issues limited the amount of visible detail. Therefore, we collected original pieces from *Teen Ink*'s physical archives. Occasionally, an original artwork was missing. In such a case, we randomly selected substitute pieces from the available originals in the archive. In total, approximately 20 percent of the early sample was chosen from this reselection process. Due to the high printing quality of the magazine issues published after September 1999, we were able to code artworks from the later period directly from the printed *Teen Ink* issues.

Two research assistants with formal arts training devised the coding scheme used to analyze the visual art. They drew on the themes that emerged from the educator interviews and focus groups, as well as formal elements of technique and interpretation in the visual arts, such as background, composition, and medium. The final coding scheme included eighteen codes in total.

To ensure consistency and accuracy in interpretation, the coding process followed a coder/shadow coder approach. Each researcher served as the primary coder for half of the pieces, while the other researcher, as the shadow coder, reviewed the decisions made by the primary coder. The two coders discussed any divergent interpretations (which were few in number) and updated the code classification after achieving consensus.

The coding was entered into the qualitative software program NVivo 9, which enabled researchers to identify trends across pieces from the early (1990–1995) and late (2006–2011) periods. These trends were documented in a series of coding reports, one for each of the eighteen codes in the coding scheme.

High School Fiction

Our high school creative sample included fifty fiction pieces written by high school students attending a school with a strong creative writing program in New Orleans. Half of the stories in our sample (n = 25) were written in 1990–1995, and half of the stories were written in 2006–2011. All stories were published in annual issues of the school's literary magazine, which selects stories for publication through a peer-review process.

Drawing stories from the same school ensured that the population of students remained relatively constant throughout the twenty-year period that comprised the focus of our investigation. Our sample initially included all of the short stories published during the two periods of interest. However, in order to avoid having a sample composed heavily of repeat authors, we ultimately excluded a number of stories written by the same student.

Two research assistants with training in English literature and composition developed the coding scheme used to analyze the high school fiction writing. As with the visual art analysis, the researchers drew on the themes that emerged from our educator interviews and focus groups, as well as technical elements in fiction writing, such as plot, setting, characters, and narrative structure. The final coding scheme included twenty-two codes in total. Like the visual art analysis, the coding process for the high school fiction employed a coder/shadow coder approach. Discrepancies in coding were rare (<1 per story) and resolved through discussion between the coders. NVivo 9 was used to explore trends across stories from the early (1990–1995) and late (2006–2011) periods.

Middle School Fiction

Our middle school creative writing sample comprised forty-four stories written by seventh- and eighth-grade students attending an independent K–8 school in Maine. Half of the stories in our sample (n = 22) were written in 1995–1998, and half of the stories were written in 2007–2009. As with our high school writing sample, all stories were published in issues of the school's literary magazine. The stories published in the magazine were not juried; the published stories are, therefore, a genuine representation of the range of work that students submitted.

The sample initially included all of the stories published during our time periods of interest: 1995–1998 and 2007–2009. If a student had multiple pieces published, we included the latest work in our sample and discounted the others. Since some issues included a greater number of stories than others, after we excluded pieces with duplicate authors, we assigned a number to each story and used a random number generator to select an approximately equal number of stories from each period.

The data analytic approach for the middle school fiction stories mirrored directly the process used for coding and analyzing the high school fiction pieces.

IV. THE GOODPLAY PROJECT

In 2008–2010, we conducted in-depth, qualitative interviews with 103 youth aged ten to twenty-five living in the greater Boston area. Funded by the MacArthur Foundation's Digital Media and Learning Initiative, this investigation focused on the ethical dimensions of young people's digital media activities, including youth's experiences with and thoughts about the "thorny" situations they've encountered online.

The middle and high school participants were recruited from one suburban and two urban public school districts, while the college-age participants were recruited from two- and four-year private and public colleges. Our research team recruited postcollege participants using Craigslist and by posting flyers in the same geographic areas as the student participants. These recruitment efforts yielded a socioeconomically and racially diverse sample.

In the interviews, participants were asked a series of questions about their experiences with digital media and how they have responded to challenging situations online, such as witnessing hate speech or deciding how much personal information to share. We also posed to them a series of hypothetical dilemmas involving ethically charged issues that are salient online, such as privacy, ownership and authorship, and community participation.

All interviews were audio recorded and transcribed verbatim. The research team devised a coding scheme that contained etic codes, identified a priori from our research questions and review of prior scholarship, and emic codes, which reflected themes that emerged directly from our line-by-line reading of the transcripts.

To ensure that researchers applied the coding scheme consistently and accurately, each researcher coded a subset of transcripts separately and then met to discuss areas of disagreement. In these meetings, the research team clarified code definitions and resolved areas of disagreement. This process was repeated until satisfactory levels of inter-coder reliability were reached for each code. Afterward, researchers divided up the codes evenly among them and applied their group of codes to the remaining transcripts. Periodic coding meetings were held throughout the coding process to ensure that researchers continued to apply their codes reliably.

After coding was complete, researchers produced a series of coding reports that summarized the patterns across participants. Also included in the reports were representative quotes that illuminated key themes.

V. TEEN BLOGGER STUDY

In 2007, Katie interviewed twenty girls who had each been blogging throughout their middle and high school years in a popular online journaling community called LiveJournal. Participants were between the ages of seventeen and twenty-one and represented all school years between tenth grade of high school and senior year of college. The majority were White (12), with a sizable minority identifying as Asian (5). The remaining girls identified themselves as Hispanic (1), Pacific Islander (1), and a mix of Native American, Black, and White (1). All participants were either residents of or attending college in the greater Boston area during the time of the study.

Each participant took part in a face-to-face interview that lasted approximately 60 minutes and followed a semistructured protocol. The questions focused on how the participants used

their blog to express themselves, explore their personal interests, and connect to other people. Participants were also asked
about how their blog had changed over the years, as well as
the relationship between their blogging and other digital media
activities like social networking, texting, and instant messaging.
These questions elicited responses that touched on the themes
of identity, intimacy, and imagination that we explore in this
book.

Following the analytic approach used by the GoodPlay team,
the interviews were audio-recorded and transcribed verbatim.
The transcripts were then coded using a coding scheme that included both etic and emic codes. To establish inter-coder agreement, two members of the GoodPlay team used the coding scheme
to code one of the transcripts. They discussed discrepancies in
their code application, and Katie used this discussion to guide
her subsequent coding. To identify themes within and across participants, Katie produced analytic memos for each participant
and conducted analyses using the qualitative software package
NVivo 8.

VI. BERMUDA STUDY

In 2010, Katie conducted a mixed-method study in Bermuda's
secondary schools (grades 8–12). The first phase of the study
involved a large-scale survey of 2,079 students (57 percent fe-
male) between the ages of eleven and nineteen years ($M = 15.4$
years) attending public and private schools in Bermuda. With
approximately 2,600 students attending secondary school in Bermuda, the survey sample accounted for roughly 80 percent of all
secondary students on the island. The second phase of the study
(whose findings we draw on throughout this book) comprised
in-depth interviews with 32 students who had participated in the

survey. The interview participants spanned the same schools and grade levels as the survey participants.

The interviews took place during school hours and followed a semistructured protocol that allowed the interviewer to explore unanticipated responses and topics that were of particular interest to participants. All participants were asked questions about their technology ownership and digital media activities, including how often they engage in these activities and their motivations for doing so. They were asked about a wide range of digital media pursuits, including social networking, texting, instant messaging, and gaming. The interview also included questions regarding the quality of participants' relationships with their friends, parents, and teachers; their experiences in school; and their self-understanding.

The analytic approach for the interviews mirrored the approaches taken in the GoodPlay Project and blogger study, described above. Each interview was audio-recorded and transcribed verbatim. Katie created a coding scheme that contained both etic and emic codes. In an effort to ensure that she applied the codes consistently and accurately, she enlisted a graduate student experienced in qualitative data analysis to code a subset of the transcripts. They each coded the same group of transcripts and then met to discuss areas of discrepancy. Once satisfactory levels of agreement had been reached, Katie coded the remainder of the transcripts.

During the coding process, Katie produced a coding memo for each participant that summarized all comments they made in relation to the codes in the coding scheme. After completing this process, she used a qualitative software program (NVivo 9) to aid in the identification of salient patterns across participants.

Notes

CHAPTER 2. TALK ABOUT TECHNOLOGY

1. Lewis Mumford, *Technics and Civilization* (New York: Harcourt Brace, 1934).

2. Jacques Ellul, *The Technological Society* (New York: Vintage, 1964).

3. Marshall McLuhan, *The Gutenberg Galaxy* (1962; reprint ed., Toronto: University of Toronto Press, 2011). Marshall McLuhan, *Understanding Media: The Extension of Man* (1964; reprint ed., Cambridge, MA: MIT Press, 1994).

4. William Wordsworth, "The French Revolution, as It Appeared to Enthusiasts" (1809).

5. Virginia Heffernan, "The Death of the Open Web," *New York Times*, May 23, 2010.

6. William James, *Habit* (1890; reprint ed., Kessinger, 2003), 66–67.

7. Y. Shoda, W. Mischel, and P. K. Peake, "Predicting Adolescent Cognitive and Self-Regulatory Competencies from Preschool Delay of Gratification: Identifying Diagnostic Conditions," *Developmental Psychology* 26 (1990): 978–986.

8. E. Bonawitz et al., "The Double-Edged Sword of Pedagogy: Instruction Limits Spontaneous Exploration and Discovery," *Cognition* 130 (2011): 322–330.

9. B. F. Skinner, *The Behavior of Organisms* (New York: Appleton Century Crofts, 1938).

10. Howard Gardner, *The Mind's New Science: A History of the Cognitive Revolution* (New York: Basic Books, 1985).

11. B. F. Skinner, *Beyond Freedom and Dignity* (New York: Knopf, 1971).

12. Mimi Ito, *Hanging Out, Messing Around, Geeking Out: Kids Living and Learning with New Media* (Cambridge, MA: MIT Press, 2009).

13. danah boyd, *A Networked Self: Identity, Community, and Culture on Social Network Sites* (New York: Routledge, 2011); Cathy N. Davidson, *Now You See It: How the Brain Science of Attention Will Transform the Way in Which We Live, Work, and Learn* (New York: Vintage, 2011); Henry Jenkins, *Convergence Culture: Where Old and New Media Collide* (New York: NYU Press, 2008); Clay Shirky, *Here Comes Everybody: The Power of Organizing without Organizations* (New York: Penguin, 2008); David Weinberger, *Too Big to Know: Rethinking Knowledge Now That the Facts Aren't the Facts, Experts Are Everywhere, and the Smartest Person in the Room Is the Room* (New York: Basic Books, 2011).

14. Nicholas Carr, *The Shallows: What the Internet Is Doing to Our Brains* (New York: Norton, 2010).

15. Mark Bauerlein, *The Dumbest Generation: How the Digital Age Stupefies Young Americans and Jeopardizes Our Future (Or, Don't Trust Anyone Under 30)* (New York: Tarcher/Penguin, 2009).

16. Cass R. Sunstein, *Going to Extremes: How Like Minds Unite and Divide* (New York: Oxford University Press, 2011).

17. Sherry Turkle, *Alone Together: Why We Expect More from Technology and Less from Each Other* (New York: Basic Books, 2011); Jaron Lanier, *You Are Not a Gadget: A Manifesto* (New York: Vintage, 2011).

CHAPTER 3. UNPACKING THE GENERATIONS

1. T. S. Eliot, *Christianity and Culture* (1948; reprint ed., Orlando, FL: Harcourt, 1976), 91.

2. Gustave Flaubert to Mlle Leroyer de Chantepie, in *The Letters of Gustave Flaubert, 1857–1880*, ed. and trans. Francis Steegmuller (Cambridge, MA: Harvard University Press, 1982), 80.

3. Gertrude Stein, quoted in Ernest Hemingway, *A Moveable Feast* (New York: Scribners, 1964), 29. Another version was used as an epigraph in Hemingway's novel *The Sun Also Rises* (1926).

4. There is an extensive sociological and historical literature on the concept of generations. Among the leading references are: Judith Burnett, *Generations: The Time Machine in Theory and Practice* (Farnham, UK: Ashgate, 2010); Glenn H. Elder Jr., John Modell, and Ross D. Parke, eds., *Children in Time and Place* (New York: Cambridge University Press, 1993); Gerhard Falk and Ursula A. Falk, *Youth Culture and the Generation G-a-p* (New York: Algora, 2005); Karl Mannheim, "The Problem of Generations," in *From Karl Mannheim,* ed. Kurt Wolff and David Kettler (London: Transaction, 1993); Katherine Newman, "Ethnography, Biography and Cultural History: Generational Paradigms in Human Development," in *Ethnography and Human Development: Context and Meaning in Social Inquiry,* ed. Richard Jessor, Anne Colby, and Richard A. Shweder (Chicago: University of Chicago Press, 1996), 371–395; and William Strauss and Neil Howe, *Generations: The History of America's Future, 1584 to 2069* (New York: William Morrow, 1991).

5. David Riesman, Nathan Glazer, and Reuel Denney, *The Lonely Crowd* (New Haven: Yale University Press, 1950).

6. William H. Whyte Jr., *The Organization Man* (New York: Simon and Schuster, 1956); Walter Isaacson and Evan Thomas, *The Wise Men: Six Friends and the World They Made* (New York: Simon and Schuster, 1986); Kenneth Keniston, *The Uncommitted: Alien Youth in American Society* (New York: Harcourt Brace and World, 1965); C. Wright Mills, *The Power Elite* (New York: Oxford University Press, 1956).

7. Erik H. Erikson, *Childhood and Society* (New York: W. W. Norton, 1950). See also Lawrence J. Friedman, *Identity's Architect: A Biography of Erik H. Erickson* (New York: Scribner, 1999).

8. Arthur Miller, *Death of a Salesman* (1949; reprint ed., New York: Penguin, 1976), 54.

9. Arthur Levine, *When Dreams and Heroes Died: A Portrait of Today's College Student* (San Francisco: Jossey-Bass, 1980); Arthur Levine and Jeanette S. Cureton, *When Hope and Fear Collide* (San Francisco: Jossey-Bass, 1998); Arthur Levine and Diane R. Dean, *Generation on a Tightrope: A Portrait of Today's College Student* (New York: John Wiley and Sons, 2012). On difficulties in forming enduring relations in

the digital era, see Sherry Turkle, *Alone Together: Why We Expect More from Technology and Less from Each Other* (New York: Basic Books, 2011).

10. Jeffrey Jensen Arnett, *Emerging Adulthood: The Winding Road from Late Teens through the Twenties* (New York: Oxford University Press, 2004).

11. Prediction attributed to Thomas J. Watson Jr., president of IBM, either 1943 or 1958 (sources disagree).

CHAPTER 4. PERSONAL IDENTITY IN THE AGE OF THE APP

1. Sherry Turkle, *Life on the Screen: Identity in the Age of the Internet* (New York: Simon and Schuster, 1995).

2. Noelle J. Hum et al., "A Picture Is Worth a Thousand Words: A Content Analysis of Facebook Profile Photographs," *Computers in Human Behavior* 27 (2011): 1828–1833; A. Moreau et al., "L'usage de Facebook et les enjeux de l'adolescence: Une étude qualitative," *Neuropsychiatrie de l'Enfance et de l'Adolescence* 60 (2012): 429–434; J. V. Peluchette and K. Karl, "Examining Students' Intended Image on Facebook: 'What Were They Thinking?!'" *Journal of Education for Business* 85 (2010): 30–37; Susannah Stern, "Producing Sites, Exploring Identities: Youth Online Authorship," in *Youth, Identity, and Digital Media*, ed. David Buckingham (Cambridge, MA: MIT Press, 2007), 95–117; Shanyang Zhao, Sherri Grasmuck, and Jason Martin, "Identity Construction on Facebook: Digital Empowerment in Anchored Relationships," *Computers in Human Behavior* 24 (2008): 1816–1836.

3. Katie Davis and Carrie James, "Tweens' Conceptions of Privacy Online: Implications for Educators," *Learning, Media and Technology* 38 (2013): 4–25.

4. Josh Miller, "What the Tech World Looks Like to a Teen," *Buzz-Feed,* January 2, 2013, http://www.buzzfeed.com/joshmiller/what-the-tech-world-looks-like-to-a-teen.

5. Erik H. Erikson, *Identity: Youth and Crisis* (New York: W. W. Norton, 1968).

6. Arthur Levine and Diane R. Dean, *Generation on a Tightrope: A Portrait of Today's College Student* (San Francisco: Jossey-Bass, 2012).

7. Tim Clydesdale, *The First Year Out: Understanding American Teens after High School* (Chicago: University of Chicago Press, 2007).

8. John H. Pryor et al., "The American Freshman: Forty Year Trends," Cooperative Institutional Research Program, Higher Education Research Institute, UCLA, 2007; John H. Pryor et al., "The American Freshman: National Norms Fall 2012," Cooperative Institutional Research Program, Higher Education Research Institute, UCLA, 2012.

9. Robert D. Putnam, *Bowling Alone: The Collapse and Revival of American Community* (New York: Simon and Schuster, 2001).

10. Alan Wolfe, *Moral Freedom: The Impossible Idea that Defines the Way We Live Now* (New York: W. W. Norton, 2001).

11. Yalda T. Uhls and Patricia M. Greenfield, "The Rise of Fame: An Historical Content Analysis," *Cyberpsychology: Journal of Psychosocial Research on Cyberspace* 5 (2011).

12. Jean M. Twenge and Joshua D. Foster, "Birth Cohort Increases in Narcissistic Personality Traits among American College Students, 1982–2009," *Social Psychological and Personality Science* 1 (2010): 99–106.

13. Jean M. Twenge and W. Keith Campbell, "Increases in Positive Self-Views among High School Students," *Psychological Science* 19 (2008): 1082–1086; Christopher Lasch, *The Culture of Narcissism* (New York: Alfred A. Knopf, 1979).

14. Levine and Dean, *Generation on a Tightrope;* Eric Greenberg with Karl Weber, *Generation We: How Millennial Youth Are Taking over America and Changing Our World Forever* (Emeryville, CA: Pachatusan, 2008).

15. Uhls and Greenfield, "Rise of Fame."

16. Jake Halpern, *Fame Junkies: The Hidden Truths behind America's Favorite Addiction* (Boston: Houghton Mifflin, 2007).

17. Mary Helen Immordino-Yang, Joanna A. Christodoulou, and Vanessa Singh, "Rest Is Not Idleness: Implications of the Brain's Default Mode for Human Development and Education," *Perspectives on Psychological Science* 7 (2012): 352–364; David M. Levy, "Information, Silence, and Sanctuary," *Ethics and Information Technology* 9 (2007): 233–236; D. M. Levy et al., "The Effects of Mindfulness Meditation Training on Multitasking in a High-Stress Information Environment,"

in *Proceedings of Graphics Interface Conference 2012* (Toronto: Canadian Information Processing Society, 2012), 45–52; Gaëlle Desbordes et al., "Effects of Mindful-Attention and Compassion Meditation Training on Amygdala Response to Emotional Stimuli in an Ordinary, Non-Meditative State," *Frontiers in Human Neuroscience* 6 (2012): 292.

18. Erikson, *Identity;* J. Bruner and D. A. Kalmar, "Narrative and Metanarrative in the Construction of Self," in *Self-Awareness: Its Nature and Development,* ed. Michael Ferrari and Robert J. Sternberg (New York: Guilford, 1998), 308–331.

19. Levy, "Information, Silence, and Sanctuary."

20. Shawn M. Bergman et al., "Millennials, Narcissism, and Social Networking: What Narcissists Do on Social Networking Sites and Why," *Personality and Individual Differences* 50 (2011): 706–711; Laura E. Buffardi and W. Keith Campbell, "Narcissism and Social Networking Web Sites," *Personality and Social Psychology Bulletin* 34 (2008): 1303–1314; Christopher J. Carpenter, "Narcissism on Facebook: Self-Promotional and Anti-Social Behavior," *Personality and Individual Differences* 52 (2012): 482–486; C. Nathan DeWall et al., "Narcissism and Implicit Attention Seeking: Evidence from Linguistic Analyses of Social Networking and Online Presentation," *Personality and Individual Differences* 51 (2011): 57–62; Bruce C. McKinney, Lynne Kelly, and Robert L. Duran, "Narcissism or Openness?: College Students' Use of Facebook and Twitter," *Communication Research Reports* 29 (2012): 108–118; Eileen Y. L. Onge et al., "Narcissism, Extraversion and Adolescents' Self-Presentation on Facebook," *Personality and Individual Differences* 50 (2011): 180–185.

21. Buffardi and Campbell, "Narcissism and Social Networking Web Sites."

22. McKinney, Kelly, and Duran, "Narcissism or Openness?"

23. Frank Rose, "The Selfish Meme," *Atlantic,* October 2012.

24. Sherry Turkle, *Alone Together: Why We Expect More from Technology and Less from Each Other* (New York: Basic Books, 2011), 268.

25. Pryor et al., "American Freshman."

26. Levine and Dean, *Generation on a Tightrope;* Sara H. Konrath, Edward H. O'Brien, and Courtney Hsing, "Changes in Dispositional Empathy in American College Students over Time: A Meta-Analysis," *Personality and Social Psychology Review* 15 (2011): 180–198.

27. Todd G. Buchholz and Victoria Buchholz, "The Go-Nowhere Generation," *New York Times*, March 10, 2012.

28. Kim Parker, "The Boomerang Generation: Feeling OK about Living with Mom and Dad," Pew Research Center, March 15, 2012, http://www.pewsocialtrends.org/2012/03/15/the-boomerang-generation/.

29. "Percentage of Teen Drivers Continues to Drop," University of Michigan News Service, July 23, 2013, http://www.ns.umich.edu/new/releases/20646-percentage-of-teen-drivers-continues-to-drop.

30. Buchholz and Buchholz, "Go-Nowhere Generation."

31. This statement echoes the argument made by Madeline Levine in an August 4, 2012, *New York Times* opinion piece. Says Levine, if you aren't willing to allow your children to be unhappy, you should not be in the parenting business.

32. "Percentage of Teen Drivers Continues to Drop."

33. Andrew R. Schrock and danah boyd, "Problematic Youth Interactions Online: Solicitation, Harassment, and Cyberbullying," in *Computer-Mediated Communication in Personal Relationships*, ed. Kevin B. Wright and Lynne M. Webb (New York: Peter Lang, 2011), 368–396.

34. Kim Thomas, "Teen Online Safety and Digital Reputation Survey," Cox Communications in partnership with the National Center for Missing and Exploited Children, June 2010, http://multivu.prnewswire.com/player/44526-cox-teen-summit-internet-safety/docs/44526-Cox_Online_Safety_Digital_Reputation_Survey-FNL.pdf.

35. Davis and James, "Tweens' Conceptions of Privacy Online."

36. Alice E. Marwick, Diego Murgia-Diaz, and John G. Palfrey Jr., *Youth, Privacy, and Reputation: Literature Review*, Berkman Center Research Publication No. 2010-5; Harvard Public Law Working Paper No. 10-29; http://ssrn.com/abstract=1588163; Kirsty Young, "Identity Creation and Online Social Networking: An Australian Perspective," *International Journal of Emerging Technologies and Society* 7 (2009): 39–57.

37. B. Ridout, A. Campbell, and L. Ellis, "Off Your Face(book)?: Alcohol in Online Social Identity Construction and Its Relation to Problem Drinking in University Students," *Drug and Alcohol Review* 31 (2012): 20–26.

38. Doug Gross, "Snapchat: Sexting Tool, or the Next Instagram?" *CNN*, January 10, 2013, http://www.cnn.com/2013/01/03/tech/mobile/snapchat/index.html.

39. Kashmir Hill, "Snapchat Won't Protect You from Jerks," *Forbes,* March 18, 2013, http://www.forbes.com/sites/kashmirhill/2013/03/18/snapchat-wont-protect-you-from-jerks/.

40. Lynn Schofield Clark, *The Parent App: Understanding Families in the Digital Age* (New York: Oxford University Press, 2013); Barbara K. Hofer and Abigail Sullivan Moore, *The iConnected Parent: Staying Close to Your Kids in College (and beyond) while Letting Them Grow Up* (New York: Free Press, 2010); Margaret K. Nelson, *Parenting Out of Control: Anxious Parents in Uncertain Times* (New York: NYU Press, 2010); Katie Davis, "A Life in Bits and Bytes: A Portrait of a College Student and Her Life with Digital Media," *Teachers College Record* 113 (2011): 1960–1982.

41. Hofer and Moore, *iConnected Parent.*

42. Hofer and Moore, *iConnected Parent.*

43. William Merrin, "MySpace and Legendary Psychasthenia," Media Studies 2.0, September 14, 2007, http://mediastudies2pointo.blogspot.com/2007/09/myspace-and-legendary-psychasthenia.html. To explain how people's online presence can weaken their sense of self to the point of full renunciation, media studies scholar William Merrin draws on Roger Callois's work around insect mimicry and psychasthenia. In 1935, Callois coined the word *psychasthenia* to describe a disorder marked by a person's inability to distinguish between oneself and one's surroundings.

According to Callois, a sense of self requires an understanding of how we relate to and are distinct from our environment. Without this ability, a sense of self cannot exist. Merrin makes the connection to cyberspace by observing that a person's online self is disconnected from the point in space he or she occupies offline. Merrin argues that we lose ourselves further as we populate predetermined identity templates on social networking sites like Facebook. Though it may feel as if we're expressing our individuality by displaying a unique combination of pictures, friends, and "likes," collectively, user profiles do more to signal the environment of Facebook than our individual personalities. Like insects that mimic their environments, we assimilate to these online environments and in so doing renounce ourselves.

44. Pryor et al., "American Freshman."

45. Levine and Dean, *Generation on a Tightrope*.

46. Eli Pariser, *The Filter Bubble: What the Internet Is Hiding from You* (New York: Penguin, 2011); Markus Prior, *Post-Broadcast Democracy: How Media Choice Increases Inequality in Political Involvement and Polarizes Elections* (New York: Cambridge University Press, 2007); Bill Bishop, *The Big Sort: Why the Clustering of Like-Minded America Is Tearing Us Apart* (Boston: Houghton Mifflin, 2008); Cass Sunstein, *Republic.com* (Princeton, NJ: Princeton University Press, 2001); Lada Adamic and Natalie Glance, "The Political Blogosphere and the 2004 U.S. Election: Divided They Blog" (Paper presented at the proceedings of WWW-2005, Chiba, Japan, May 2005); J. Kelly, D. Fisher, and M. Smith, "Debate, Division, and Diversity: Political Discourse Networks in USENET Newsgroups" (Paper presented at the Second Conference on Online Deliberation: Design, Research, and Practice [DIAC 05], Stanford University, Palo Alto, CA, May 2005).

For counterevidence, see Matthew Gentzkow and Jesse M. Shapiro, "Ideological Segregation Online and Offline," *Quarterly Journal of Economics* 126 (2011): 1799–1839. And for a mix of confirmatory and counterevidence, see Sarita Yardi and danah boyd, "Dynamic Debates: An Analysis of Group Polarization over Time on Twitter," *Bulletin of Science, Technology and Society* 30 (2011): 316–327; Eszter Hargittai, Jason Gallo, and Matthew Kane, "Cross-Ideological Discussions among Conservatives and Liberal Bloggers," *Public Choice* 134 (2008): 67–86; H. Farrell, "The Consequences of the Internet for Politics," *Annual Review of Political Science* 15 (2012): 35–52; and Itai Himelboim, Stephen McCreery, and Marc Smith, "Birds of a Feather Tweet Together: Integrating Network and Content Analyses to Examine Cross-Ideology Exposure on Twitter," *Journal of Computer-Mediated Communication* 18 (2013): 40–60.

47. Clay Shirky, *Here Comes Everybody: The Power of Organizing without Organizations* (New York: Penguin, 2008).

48. Y. Benkler, *The Wealth of Networks: How Social Production Transforms Markets and Freedom* (New Haven: Yale University Press, 2006).

49. Mimi Ito et al., *Hanging Out, Messing Around, and Geeking Out: Kids Living and Learning with New Media* (Cambridge, MA: MIT Press, 2009).

CHAPTER 5. APPS AND INTIMATE RELATIONSHIPS

1. K. Davis, "Friendship 2.0: Adolescents' Experiences of Belonging and Self-Disclosure Online," *Journal of Adolescence* 35 (2012): 1527–1536; Yochai Benkler, *The Wealth of Networks: How Social Production Transforms Markets and Freedom* (New Haven: Yale University Press, 2006); Mimi Ito et al., *Hanging Out, Messing Around, and Geeking Out: Kids Living and Learning with New Media* (Cambridge, MA: MIT Press, 2009).

2. Mary Madden et al., "Teens and Technology 2013," Pew Internet and American Life Project, March 13, 2013, available at: http://www.pewinternet.org/~/media//Files/Reports/2013/PIP_TeensandTechnology 2013.pdf.

3. Amanda Lenhart, "Teens, Smartphones, and Texting," Pew Internet and American Life Project, March 19, 2012, http://pewinternet.org/Reports/2012/Teens-and-smartphones.aspx.

4. Davis, "Friendship 2.0."

5. R. Ling, and B. Yttri, "Hyper-Coordination via Mobile Phones in Norway," in *Perpetual Contact: Mobile Communication, Private Talk, Public Performance,* ed. J. E. Katz and M. Aakhus (Cambridge: Cambridge University Press, 2006), 139–169.

6. Mimi Ito and Daisuke Okabe, "Technosocial Situations: Emergent Structuring of Mobile E-Mail Use," in *Personal, Portable, Pedestrian: Mobile Phones in Japanese Life,* ed. M. Ito, D. Okabe, and M. Matsuda (Cambridge, MA: MIT Press, 2005), 257–273.

7. Barbara K. Hofer and Abigail Sullivan Moore, *The iConnected Parent: Staying Close to Your Kids in College (and beyond) while Letting Them Grow Up* (New York: Free Press, 2010); Margaret K. Nelson, *Parenting Out of Control: Anxious Parents in Uncertain Times* (New York: NYU Press, 2010).

8. danah boyd, "Why Youth Heart Social Network Sites: The Role of Networked Publics in Teenage Social Life," in *Youth, Identity, and Digital Media,* ed. David Buckingham (Cambridge, MA: MIT Press, 2007), 119–142.

9. Joseph B. Walther, "Computer-Mediated Communication: Impersonal, Interpersonal, and Hyperpersonal Interaction," *Communication Research* 23 (1996): 3–43. Walther's hyperpersonal communication

theory states that specific features of computer-mediated communication, such as audiovisual anonymity and asynchrony, encourage people to self-disclose more than they would through face-to-face communication.

10. Luigi Bonetti, Marilyn Anne Campbell, and Linda Gilmore, "The Relationship of Loneliness and Social Anxiety with Children's and Adolescents' Online Communication," *Cyberpsychology, Behavior, and Social Networking* 13 (2010): 279–285; Alexander P. Schouten, Patti M. Valkenburg, and Jochen Peter, "Precursors and Underlying Processes of Adolescents' Online Self-Disclosure: Developing and Testing an 'Internet-Attribute-Perception' Model," *Media Psychology* 10 (2007): 292–315; Susannah Stern, "Producing Sites, Exploring Identities: Youth Online Authorship," in Buckingham, *Youth, Identity, and Digital Media*, 95–117; Patti M. Valkenburg and Jochen Peter, "Social Consequences of the Internet for Adolescents: A Decade of Research," *Current Directions in Psychological Science* 18 (2009): 1–5; Patti M. Valkenburg and Jochen Peter, "Online Communication among Adolescents: An Integrated Model of Its Attraction, Opportunities, and Risks," *Journal of Adolescent Health* 48 (2011): 121–127; P. M. Valkenburg, S. R. Sumter, and J. Peter, "Gender Differences in Online and Offline Self-Disclosure in Pre-Adolescence and Adolescence," *British Journal of Developmental Psychology* 29 (2011): 253–269.

11. Miller McPherson, Lynn Smith-Lovin, and Matthew E. Brashears, "Social Isolation in America: Changes in Core Discussion Networks over Two Decades," *American Sociological Review* 71 (2006): 353–375. Though published in a highly respected, peer-reviewed journal, McPherson's findings have not gone without challenge. Another sociologist, Claude Fischer of UC Berkeley, believes that the findings are an artifact of the 2004 survey process, which he argues contained a number of anomalies and inconsistencies. He cautions: "Scholars and general readers alike should draw no inference from the 2004 GSS as to whether Americans' social networks changed substantially between 1985 and 2004; they probably did not." Though McPherson and his colleagues published a convincing rebuttal, we believe it's still important to acknowledge that their findings haven't convinced everyone.

In addition, a 2009 study conducted by the Pew Internet and American Life Project concluded that Americans may not be quite as socially isolated as the McPherson et al. study suggests, though the study did confirm that Americans' discussion networks have become smaller and less diverse since 1985: Keith Hampton et al., "Social Isolation and New Technology," November 4, 2009, http://pewinternet.org/Reports/2009/18—Social-Isolation-and-New-Technology.aspx. In Europe, Leopoldina Fortunati and colleagues found that the percentage of Europeans who regularly visit friends and relatives decreased between 1996 and 2009. Other forms of in-person sociability, such as taking part in sporting activities and going out to restaurants, pubs, or dancing, became less frequent during the same time period (though a greater percentage of Europeans reported taking part in such activities). Interestingly, those with access to the Internet were more likely to report taking part in various forms of in-person sociability. Leopoldina Fortunati, Sakari Taipale, and Federico de Luca, "What Happened to Body-to-Body Sociability?" *Social Science Research* 42 (2013): 893–905.

12. National Opinion Research Center, *The General Social Survey (GSS), 1972–2008* (data file), accessed in 2009, http://www.norc.org; R. D. Putman, *Bowling Alone: The Collapse and Revival of American Community* (New York: Simon and Schuster, 2000); R. V. Robinson, and E. F. Jackson, "Is Trust in Others Declining in America? An Age-Period-Cohort Analysis," *Social Science Research* 30 (2001): 117–145; Katie Davis et al., "'I'll Pay Attention When I'm Older: Generational Differences in Trust," in *Restoring Trust in Organizations and Leaders: Enduring Challenges and Emerging Answers,* ed. Roderick M. Kramer and Todd L. Pittinsky (New York: Oxford University Press, 2012), 47–67; Katie Davis and Howard Gardner, "Trust: Its Conceptualization by Scholars, Its Status with Young Persons," in *Political and Civic Leadership: A Reference Handbook,* ed. Richard A. Couto, vol. 2 (Thousand Oaks, CA: Sage, 2010), 602–610.

13. Sherry Turkle, *Alone Together: Why We Expect More from Technology and Less from Each Other* (New York: Basic Books, 2011); Jacqueline Olds and Richard Schwartz, *The Lonely America;* Stephen Marche, "Is Facebook Making Us Lonely?" *Atlantic,* May 2012; David DiSalvo, "Are Social Networks Messing with Your Head?" *Scientific American,* January–February 2010.

14. R. Pea et al., "Media Use, Face-to-Face Communication, Media Multitasking, and Social Well-Being among 8–12-Year-Old Girls," *Developmental Psychology* 48 (2012): 327–336.

15. Hui-Tzu Grace Chou and Nicholas Edge, "'They Are Happier and Having Better Lives than I Am': The Impact of Facebook on Perceptions of Others' Lives," *Cyberpsychology, Behavior, and Social Networking* 15 (2012): 117–121.

16. Odelia Kaly, "Why I'm Worried about Social Media," *Huffington Post*, http://www.huffingtonpost.com/odelia-kaly/why-im-worried -about-soci_b_2161554.html.

17. Turkle, *Alone Together*; Andrew Reiner, "Only Disconnect," *Chronicle of Higher Education*, September 24, 2012.

18. Caroline Tell, "Let Your Smartphone Deliver the Bad News," *New York Times*, October 26, 2012.

19. Turkle, *Alone Together*, 154.

20. Christy Wampole, "How to Live without Irony," *New York Times*, November 17, 2012, http://opinionator.blogs.nytimes.com/2012/ 11/17/how-to-live-without-irony/.

21. Hofer and Moore, *iConnected Parent*.

22. Amanda L. Williams and Michael J. Merten, "iFamily: Internet and Social Media Technology in the Family Context," *Family and Consumer Sciences Research Journal* 40 (2011): 150–170.

23. Barry Wellman et al., "Connected Lives: The Project," in *Networked Neighbourhoods: The Connected Community in Context*, ed. Patrick Purcell (Berlin: Springer, 2005), 161–216.

24. Valkenburg and Peter, "Social Consequences of the Internet for Adolescents"; Valkenburg and Peter, "Online Communication among Adolescents"; Nicole B. Ellison, Charles Steinfield, and Cliff Lampe, "The Benefits of Facebook 'Friends': Social Capital and College Students' Use of Online Social Network Sites," *Journal of Computer-Mediated Communication* 12 (2007): 1143–1168; Keith N. Hampton, Lauren F. Sessions, and Eun Ja Her, "Core Networks, Social Isolation, and New Media: How Internet and Mobile Phone Use Is Related to Network Size and Diversity," *Information Communication and Society* 14 (2011): 130–155; Hua Wang and Barry Wellman, "Social Connectivity in America: Changes in Adult Friendship Network Size from 2002 to 2007," *American Behavioral Scientist* 53 (2010): 1148–1169; Ito et al., *Hanging Out, Messing Around, and Geeking Out*; S. Craig Watkins, *The Young and the Digital: What the Migration to Social Network Sites, Games, and Anytime, Anywhere Media Means for Our Future* (Boston:

Beacon, 2010); Lee Rainie and Barry Wellman, *Networked: The New Social Operating System* (Cambridge, MA: MIT Press, 2012).

25. Davis, "Friendship 2.0."

26. Susannah Stern, "Producing Sites, Exploring Identities: Youth Online Authorship," in Buckingham, *Youth, Identity, and Digital Media,* 95–117.

27. Sara H. Konrath, Edward H. O'Brien, and Courtney Hsing, "Changes in Dispositional Empathy in American College Students over Time: A Meta-Analysis," *Personality and Social Psychology Review* 15 (2011): 180–198; Ito et al., *Hanging Out, Messing Around, and Geeking Out.*

28. Arthur Levine and Diane R. Dean, *Generation on a Tightrope: A Portrait of Today's College Student* (San Francisco: Jossey-Bass, 2012).

29. Associated Press–MTV Digital Abuse Survey, http://surveys.ap .org/data%5CKnowledgeNetworks%5CAP_DigitalAbuseSurvey_Top lineTREND_1st%20story.pdf.

30. Amanda Lenhart et al., "Teens, Kindness, and Cruelty on Social Network Sites," *Pew Internet and American Life Project,* November 9, 2011, http://pewinternet.org/Reports/2011/Teens-and-social-media.aspx.

31. For an in-depth analysis of the role that social media play in bullying among teens, see Emily Bazelon, *Sticks and Stones: Defeating the Culture of Bullying and Rediscovering the Power of Character and Empathy* (New York: Random House, 2013).

32. Amy O'Leary, "In Virtual Play, Sex Harassment Is All Too Real," *New York Times,* August 1, 2012.

Though many games and gaming communities appear to encourage aggressive, demeaning behavior toward others, we're encouraged by the emergence of educational games that promote prosocial and ethical behavior. See, e.g., T. Greitemeyer and S. Osswald, "Effects of Prosocial Video Games on Prosocial Behavior," *Journal of Personality and Social Psychology* 98 (2010): 211–221; and Marc A. Sestir and Bruce D. Bartholow, "Violent and Nonviolent Video Games Produce Opposing Effects on Aggressive and Prosocial Outcomes," *Journal of Experimental Social Psychology* 46 (2010): 934–942.

33. Caroline Heldman and Lisa Wade, "Hook-Up Culture: Setting a New Research Agenda," *Sexual Research and Social Policy* 7 (2010): 323–333; Donna Freitas, *The End of Sex: How Hookup Cul-*

ture Is Leaving a Generation Unhappy, Sexually Unfulfilled, and Confused about Intimacy (New York: Basic Books, 2013); J. R. Garcia and C. Reiber, "Hook-Up Behavior: A Biopsychosocial Perspective," *Journal of Social, Evolutionary, and Cultural Psychology* 2 (2008): 49–65; Madeline A. Fugère et al., "Sexual Attitudes and Double Standards: A Literature Review Focusing on Participant Gender and Ethnic Background," *Sexuality and Culture* 12 (2008): 169–182.

34. Levine and Dean, *Generation on a Tightrope.*

35. Ajay T. Abraham, Anastasiya Pocheptsova, and Rosellina Ferraro, "The Effect of Mobile Phone Use on Prosocial Behavior," http://gfx.svd -cdn.se/multimedia/archive/00830/L_s_hela_studien_om_830163a.pdf.

36. Eli Pariser, *The Filter Bubble: What the Internet Is Hiding from You* (New York: Penguin, 2011); Markus Prior, *Post-Broadcast Democracy: How Media Choice Increases Inequality in Political Involvement and Polarizes Elections* (New York: Cambridge University Press, 2007); Bill Bishop, *The Big Sort* (Boston: Houghton Mifflin, 2008); Cass R. Sunstein, *Republic.com* (Princeton, NJ: Princeton University Press, 2001); Lada Adamic and Natalie Glance, "The Political Blogosphere and the 2004 U.S. Election: Divided They Blog" (Paper presented at the proceedings of WWW-2005, Chiba, Japan, May 2005); J. Kelly, D. Fisher, and M. Smith, "Debate, Division, and Diversity: Political Discourse Networks in USENET Newsgroups" (Paper presented at the Second Conference on Online Deliberation: Design, Research, and Practice [DIAC 05], Stanford University, Palo Alto, CA, May 2005); Itai Himelboim, Stephen McCreery, and Marc Smith, "Birds of a Feather Tweet Together: Integrating Network and Content Analyses to Examine Cross-Ideology Exposure on Twitter," *Journal of Computer-Mediated Communication* 18 (2013): 40–60.

37. For a counterargument, see Farhad Manjoo's essay, "My Technology New Year's Resolutions," *Slate,* January 4, 2013, in which he argues that Twitter is the "anti-filter bubble"; http://www.slate.com/ articles/technology/technology/2013/01/new_year_s_resolutions_for _technology_in_2013.html. For counterevidence, see Matthew Gentzkow and Jesse M. Shapiro, "Ideological Segregation Online and Offline," *Quarterly Journal of Economics* 126 (2011): 1799–1839. And for a mix of confirmatory and counter-evidence, see Sarita Yardi and danah boyd, "Dynamic Debates: An Analysis of Group Polarization

over Time on Twitter," *Bulletin of Science, Technology and Society* 30 (2010): 316–327; Eszter Hargittai, Jason Gallo, and Matthew Kane, "Cross-Ideological Discussions among Conservatives and Liberal Bloggers," *Public Choice* 134 (2008): 67–86; H. Farrell, "The Consequences of the Internet for Politics," *Annual Review of Political Science* 15 (2012): 35–52.

CHAPTER 6. ACTS (AND APPS) OF IMAGINATION AMONG TODAY'S YOUTH

1. Mimi Ito et al., *Hanging Out, Messing Around, and Geeking Out: Kids Living and Learning with New Media* (Cambridge, MA: MIT Press, 2009); Henry Jenkins, *Convergence Culture: Where Old and New Media Collide* (New York: NYU Press, 2006).

2. Mihaly Csikszentmihalyi, "Implications of a Systems Perspective for the Study of Creativity," in *Handbook of Creativity*, ed. Robert J. Sternberg (Cambridge: Cambridge University Press, 1999), 313–335.

3. Clay Shirky, *Cognitive Surplus: Creativity and Generosity in a Creative Age* (New York: Penguin, 2011).

4. Jenkins, *Convergence Culture*.

5. Jaron Lanier, *You Are Not a Gadget: A Manifesto* (New York: Alfred A. Knopf, 2010); Howard Gardner, *Truth, Beauty, and Goodness Reframed: Educating for the Virtues in the Age of Truthiness and Twitter* (New York: Basic Books, 2011).

6. Elizabeth Bonawitz et al., "The Double-Edged Sword of Pedagogy: Instruction Limits Spontaneous Exploration and Discovery," *Cognition* 120 (2011): 322–330.

7. Kyung Hee Kim, "The Creativity Crisis: The Decrease in Creative Thinking Scores on the Torrance Tests of Creative Thinking," *Creativity Research Journal* 23 (2011): 285–295.

8. Kyung Hee Kim, "Meta-Analyses of the Relationship of Creative Achievement to Both IQ and Divergent Thinking Test Scores," *Journal of Creative Behavior* 42 (2008): 106–130; E. Paul Torrance, "Prediction of Adult Creative Achievement among High School Seniors," *Gifted Child Quarterly* 13 (1969): 223–229; E. Paul Torrance, "Predictive Validity of the Torrance Tests of Creative Thinking," *Journal of Creative Behavior* 6 (1972): 236–252; Hiroyuki Yamada and Alice Yu-Wen Tam,

"Prediction Study of Adult Creative Achievement: Torrance's Longitudinal Study of Creativity Revisited," *Journal of Creative Behavior* 30 (1996): 144–149.

9. Kim, "Creativity Crisis"; Po Bronson and Ashley Merryman, "The Creativity Crisis," *Newsweek,* July 10, 2010, http://www.thedaily beast.com/newsweek/2010/07/10/the-creativity-crisis.html; Tom Ashbrook, "U.S. Creativity in Question," On-Point with Tom Ashbrook, WBUR, July 20, 2010, http://onpoint.wbur.org/2010/07/20/u-s-creativ ity-in-question.

10. Sandra W. Russ and Jessica A. Dillon, "Changes in Children's Pretend Play over Two Decades," *Creativity Research Journal* 23 (2011): 330–338.

11. On pretend play, see Edward P. Fisher, "The Impact of Play on Development: A Meta-Analysis," *Play and Culture* 5 (1992): 159–181; on divergent thinking, see Beth A. Hennessey and Teresa M. Amabile, "Creativity," *Annual Review of Psychology* 61 (2010): 569–598; Howard B. Parkhurst, "Confusion, Lack of Consensus, and the Definition of Creativity as a Construct," *Journal of Creative Behavior* 33 (1999): 1–21; Joy P. Guilford, *The Nature of Human Intelligence* (New York: McGraw-Hill, 1967).

12. Sandra W. Russ and Ethan D. Schafer, "Affect in Fantasy Play, Emotion in Memories, and Divergent Thinking," *Creativity Research Journal* 18 (2006): 347–354.

13. On remix culture, see Ito et al., *Hanging Out, Messing Around, and Geeking Out;* Jenkins, *Convergence Culture;* and Shirky, *Cognitive Surplus.*

14. Lanier, *You Are Not a Gadget,* 20.

15. Betsy Sparrow, Jenny Liu, and Daniel M. Wegner, "Google Effects on Memory: Cognitive Consequences of Having Information at Our Fingertips," *Science* 333 (2011): 776–778.

16. Patricia Greenfield and Jessica Beagles-Roos, "Radio vs. Television: Their Cognitive Impact on Children of Different Socioeconomic and Ethnic Groups," *Journal of Communication* 38 (1988): 71–92.

17. Patti M. Valkenburg and Tom H. A. van der Voort, "Influence of TV on Daydreaming and Creative Imagination: A Review of Research," *Psychological Bulletin* 116 (1994): 316–339.

18. Shirley Brice Heath, personal communication with author, June 3, 2011.

19. Eric Hoover, "Boston College Sees a Sharp Drop in Applications after Adding an Essay," *Boston Globe,* January 16, 2013.

20. Mihaly Csikszentmihalyi, *Creativity: Flow and the Psychology of Discovery and Invention* (New York: Harper Perennial, 1997).

21. Brenda Patoine, "Brain Development in a Hyper-Tech World," Dana Foundation, August 2008, http://www.dana.org/media/detail.aspx ?id=13126.

22. Karin Foerde, Barbara J. Knowlton, and Russell A. Poldrack, "Modulation of Competing Memory Systems by Distraction," *PNAS* 103 (2006): 11778–11783.

23. Sophie Ellwood, Gerry Pallier, Allan Snyder, and Jason Gallate, "The Incubation Effect: Hatching a Solution?" *Creativity Research Journal* 21 (2009): 6–14.

24. Flora Beeftink, Wendelien van Eerde, and Christel G. Rutte, "The Effect of Interruptions and Breaks on Insight and Impasses: Do You Need a Break Right Now?" *Creativity Research Journal* 20 (2008): 358–364.

25. Brewster Ghiselin, *The Creative Process: A Symposium* (Berkeley: University of California Press, 1952).

26. Ito et al., *Hanging Out, Messing Around, and Geeking Out;* Jacob W. Getzels and Mihaly Csikszentmihalyi, *The Creative Vision: A Longitudinal Study of Problem Finding in Art* (New York: John Wiley and Sons, 1976).

27. William Poundstone, *Are You Smart Enough to Work at Google? Trick Questions, Zen-Like Riddles, Insanely Difficult Puzzles and Other Devious Interviewing Techniques You Need to Know to Get a Job Anywhere in the New Economy* (New York: Back Bay Books/Little, Brown, 2012).

28. Seymour Papert, *Mindstorms: Children, Computers, and Powerful Ideas* (New York: Basic Books, 1980); Mitchel Resnick and Brian Silverman, "Some Reflections on Designing Construction Kits for Kids," in *IDC '05: Proceedings of the 2005 Conference on Interaction Design and Children* (New York: ACM, 2005), 117–122.

29. Linda A. Jackson et al., "Information Technology Use and Cre-

ativity: Findings from the Children and Technology Project," *Computers in Human Behavior* 28 (2012): 370–376.

30. Oscar Ardaiz-Villanueva et al., "Evaluation of Computer Tools for Idea Generation and Team Formation in Project-Based Learning," *Computers and Education* 56 (2011): 700–711.

31. Igor Stravinsky, *Poetics of Music in the Form of Six Lessons* (Cambridge, MA: Harvard University Press, 1942), 63.

32. Shirky, *Cognitive Surplus.*

33. Lawrence Lessig, *Code and Other Laws of Cyberspace* (New York: Basic Books, 2000).

34. Hennessey and Amabile, "Creativity"; Parkhurst, "Confusion, Lack of Consensus, and the Definition of Creativity as a Construct"; Guilford, *Nature of Human Intelligence.*

CHAPTER 7. CONCLUSION

Epigraph: Alfred North Whitehead, *An Introduction to Mathematics* (New York: Holt, 1911), 61. We were pleased to see that Evgeny Morozov reflected similarly on this quotation in a recent column. See Morozov, "Machines of Laughter and Forgetting," *New York Times Sunday Review,* March 31, 2013, 12.

1. Anthony Burgess, *A Clockwork Orange* (1962; reprint ed., New York: W. W. Norton, 1986).

2. Anthony Burgess, "The Clockwork Condition," *New Yorker,* June 4, 2012.

3. Burgess, "Clockwork Condition."

4. Aldous Huxley, *Brave New World* (1932; reprint ed., New York: Harcourt Perennial, 2006).

5. George Orwell, *1984* (1948; reprint ed., New York: Signet, 1961).

6. B. F. Skinner, *Walden II* (1948; reprint ed., New York: Prentice Hall, 1976); Skinner, *Beyond Freedom and Dignity* (New York: Knopf, 1972).

7. Burgess, "Clockwork Condition."

8. Gustave Flaubert, *Sentimental Education: The Story of a Young Man* (1869; reprint ed., Charleston, SC: Forgotten Books, 2012).

9. For a sampling of accounts by visitors to America, see Oscar Hand-

lin, *This Was America* (Cambridge, MA: Harvard University Press, 1949); J. Hector St. John de Crèvecoeur, *Letters from an American Farmer* (1782; reprint ed., New York: Dover, 2005); Charles Dickens, *American Notes for General Circulation* (London: Chapman and Hall, 1842); Harriet Martineau, *Society in America* (1837; reprint ed., New Brunswick, NJ: Transaction, 1981); Frances Trollope, *Domestic Manners of the Americans* (1832; reprint ed., New York: Dover, 2003); Alexis de Tocqueville, *Democracy in America* (1835, 1840; new trans., New York: Harper Perennial Classics, 2006); Alistair Cooke, *Alistair Cooke's America* (1973; reprint ed., New York: Basic Books, 2009); and D. W. Brogan, *The American Character* (New York: Alfred A. Knopf, 1944).

10. Daniel Gilbert, *Stumbling on Happiness* (New York: Random House, 2006).

11. Shirley Brice Heath made this comment at a seminar at Harvard Project Zero on March 29, 2011, and at the Annual Meeting of the National Academy of Education, October 30, 2011.

12. Jeffrey Jensen Arnett, *Emerging Adulthood: The Winding Road from the Late Teens through the Twenties* (New York: Oxford University Press, 2004).

13. Mark Twain, *The Adventures of Huckleberry Finn* (1885; reprint ed., New York: St. Martin's, 1995), 265.

14. Robert D. Putnam and David E. Campbell, *American Grace: How Religion Divides and Unites Us* (New York: Simon and Schuster, 2010). On religious apps, see Cathleen Falsani, "Need Religion? There's an App for That," *Huffington Post*, December 3, 2010, http://www.huffingtonpost.com/cathleen-falsani/need-religion-theres-an-a_b_789423.html.

15. For references on Good Play, see Carrie James et al., *Young People, Ethics, and the Digital Media: A Synthesis from the GoodPlay Project* (Cambridge: MIT Press, 2009). For details, see thegoodproject.org.

16. On digital ethics, see also Marc Prensky, *Brain Gain: Technology and the Quest for Digital Wisdom* (New York: Palgrave Macmillan, 2012).

17. Alan Wolfe, *Moral Freedom: The Search for Virtue in a World of Choice* (New York: W. W. Norton, 2002).

18. On people believing that they are well motivated, see Dan Ariely, *The (Honest) Truth about Dishonesty: How We Lie to Everyone—Especially Ourselves* (New York: HarperCollins, 2012).

19. James et al., *Young People, Ethics, and the Digital Media*. For details, see thegoodproject.org. The Good Play Project, *Our Space: Being a Responsible Citizen of the Digital World* (Project Zero, Harvard Graduate School of Education, and Annenberg School for Communication, University of Southern California, 2011), http://dmlcentral.net/sites/dml central/files/resource_files/Our_Space_full_casebook_compressed.pdf. See also the Common Sense Media digital citizenship curriculum, http:// www.commonsensemedia.org/educators/curriculum.

20. Katie Davis et al., "Fostering Cross-Generational Dialogues about the Ethics of Online Life," *Journal of Media Literacy Education* 2 (2010): 124–150.

21. Katie Davis and Howard Gardner, "Five Minds Our Children Deserve: Why They're Needed, How to Nurture Them," *Journal of Educational Controversy* 6 (2012), http://www.wce.wwu.edu/Resources/ CEP/eJournal/v006n001/a001.shtml.

22. See Michael Polanyi, *Personal Knowledge: Towards a Post-Critical Philosophy* (Chicago: University of Chicago Press, 1958). See also Jean Lave and Etienne Wenger, *Situated Learning: Legitimate Peripheral Participation* (Cambridge: Cambridge University Press, 1991).

23. For studies of creativity, see Howard Gardner, *Creating Minds* (New York: Basic Books, 1993); and Gardner, *Extraordinary Minds: Portraits of Four Exceptional Individuals and an Examination of Our Own Extraordinariness* (New York: Basic Books, 1997).

24. Pasi Sahlberg, *Finnish Lessons: What Can the World Learn from Educational Change in Finland?* (New York: Teachers College Press, 2011).

25. Atul Gawande, *The Checklist Manifesto: How to Get Things Right* (New York: Henry Holt, 2009); Jerome Groopman, MD, *How Doctors Think* (New York: Mariner Books, 2008).

26. Howard Gardner, Mihaly Csikszentmihalyi, and William Damon, *Good Work: When Excellence and Ethics Meet* (New York: Basic Books, 2001); Howard Gardner, ed., *GoodWork: Theory and Practice* (Cambridge, MA: Good Project, 2010), http://www.goodworkproject

.org/publication/goodwork-theory-and-practice/. For more information, see the GoodWork website at http://www.thegoodproject.org/.

27. Kathleen Farrell, "Taking Stock: The Value of Structuring Reflection on GoodWork," in H. Gardner, GoodWork, http://www.good workproject.org/publication/goodwork-theory-and-practice/.

28. Home page, The Partnership for 21st Century Skills, http://www .p21.org/.

29. B. F. Skinner, *The Technology of Teaching* (Englewood Cliffs, NJ: Prentice-Hall, 1968).

30. Seth Kugel, "Using TripAdvisor? Some Advice," *New York Times,* January 1, 2013, http://frugaltraveler.blogs.nytimes.com/2013/01/01/ using-tripadvisor-some-advice/.

31. Alfred North Whitehead, *The Aims of Education and Other Essays* (New York: Free Press, 1967).

32. Matthew Arnold, "Sweetness and Light," in *"Culture and Anarchy" and Other Writings,* ed. Stefan Collini (Cambridge: Cambridge University Press, 1993), 79.

33. Jennifer Pahlka, "Code American" (Paper presented at the Aspen Ideas Festival, August 2, 2012).

34. For the Boston app to identify potholes, see http://codeforamer ica.org/2011/02/23/boston-citizens-connected/.

35. Tod Machover quoted in Jeremy Eichler, "Sounds of a City: A New Template for Collaboration in Toronto," *Boston Globe,* January 26, 2013.

36. Mimi Ito, *Engineering Play: A Cultural History of Children's Software* (Cambridge, MA: MIT Press, 2009); Tim Wu, *The Master Switch: The Rise and Fall of Information Empires* (New York: Vintage, 2011).

37. On singularity, see Brian Christian, *The Most Human Human: What Artificial Intelligence Teaches Us about Being Alive* (New York: Anchor Books, 2011); Evan Goldstein, "The Strange Neuroscience of Immortality," *The Chronicle of Higher Education,* July 20, 2012; and Ray Kurzweil, *The Singularity Is Near: When Humans Transcend Biology* (New York: Penguin, 2006).

38. Christine Rosen, "The Machine and the Ghost," *New Republic,* August 2, 2012; Allen Tate, *The Forlorn Demon: Didactic and Critical Essays* (Chicago: Regnery, 1953).

Index

ABC network, 46
action and restriction, paradox of, 24–25
adolescence, 45, 53, 55, 108
The Adventures of Huckleberry Finn (Twain), 168
Affect in Play Scale (APS), 129
AIDS, 55
The Aims of Education (White-head), 186
alcohol abuse, 78, 83
Alone Together (Turkle), 77, 100
Amazon (company), 52, 192, 195
American Idol (TV show), 68
An American in Paris (Gershwin), 190
The Andy Griffith Show (TV show), 69
anonymity, 63, 170, 219n9
anxiety, 77–81
app-directed, 42

"App Generation," 6–14, 17, 119, 153, 166; disengagement from digital world, 191–92; packaged identities of, 66; psychology of users and, 54
Apple, Inc., 24, 25, 109, 192
Apple Macintosh ("Mac"), 23–24
apps (applications), 6–7, 14, 52, 152–53; app mentality or worldview, 94, 160–61, 175; creative limits of remix culture, 142–44; designers of, 60; educational, 179–80; e-readers, 58; as filters, 104; GPS (Global Positioning System), 8–9; habits and, 24–25; icons of, 72, 91; identity and, 32, 60; interfaces of, 61; intimacy and, 32–33; pervasiveness of, 160; as portals to globalized world, 89;